T0395568

"How do you refute officious government-promoted medical falsehoods and gaslighting? This book leads by example. With sardonic wit, de Becker carefully, calmly and objectively documents how government, industry and academia have colluded to create false narratives that seem definitive. Most of all, he invites readers to draw their own conclusions."

—Robert Malone, MD, MS, physician, virologist, molecular biologist, and co-chair of CDC's Advisory Committee on Immunization Practices

"In an age when truth is buried beneath mountains of marketing, propaganda, and official narratives, de Becker delivers a much-needed roadmap to reality. Through gripping historical examples, he shows that when powerful institutions are faced with inconvenient truths, they don't investigate; they 'debunk.' This provocative and rigorously referenced exposé empowers readers to evaluate original source material for themselves—distinguishing truth from manipulation, science from spin."

—Pierre Kory, MD, MPA specialist in internal medicine, pulmonary diseases, and critical care medicine

"This book is a treasure trove of information that has been widely censored for decades. It is right on the mark."

—Christiane Northrup, MD, multiple NYTimes bestselling author of *Women's Bodies, Women's Wisdom*

"Gavin de Becker has put his formidable investigative skills into this highly readable exposé of the US government's history of hiding the harms of various substances. Just as *The Gift of Fear* became a classic reference work about violent behavior, *Forbidden Facts* is destined to become a classic about government concealment."

—Peter A. McCullough, MD, co-author of *Vaccines: Mythology, Ideology, and Reality*

FORBIDDEN FACTS

Skyhorse Publishing books may be purchased in bulk at special discounts for sales promotion, corporate gifts, fund-raising, or educational purposes. Special editions can also be created to specifications. For details, contact the Special Sales Department, Skyhorse Publishing, 307 West 36th Street, 11th Floor, New York, NY 10018 or info@skyhorsepublishing.com.

Visit our website at www.skyhorsepublishing.com.

10 9 8 7 6 5 4 3 2 1

Library of Congress Cataloging-in-Publication Data is available on file.

Hardcover ISBN: 978-1-5107-8595-3
Ebook ISBN: 978-1-5107-8596-0

Cover design by Brian Peterson
Book design by Geoff Towle

Printed in the United States of America

FORBIDDEN FACTS

GOVERNMENT DECEIT & SUPPRESSION ABOUT BRAIN DAMAGE FROM CHILDHOOD VACCINES

GAVIN de BECKER

Skyhorse Publishing

If we are not able to ask skeptical questions, to interrogate those who tell us that something is true, to be skeptical of those in authority, then we're up for grabs for the next charlatan— political or religious— who comes ambling along.

It wasn't enough, Jefferson said, to enshrine some rights in a constitution or a bill of rights. The people had to be educated, and they had to practice their skepticism… otherwise, we don't run the government, the government runs us.

— Carl Sagan

Table of Contents

FORBIDDEN FACTS

by

GAVIN de BECKER

Author's Note

As you read this book, put a big X through anything you believe is not true. ***Next to that X, write what you determine is true***. Without that second step, any of us might automatically discount or disbelieve accurate information simply because it differs from what we've heard or assumed. QR codes throughout this book link directly to original source material that can be assessed for accuracy and credibility. This way, each reader can decide what's promotion, what's marketing, what's propaganda, what's plain lies, and what is truth.

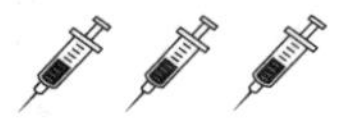

This book contains extensive favorable information about vaccines; it is not anti-vaccine or pro-vaccine. This book is about government deceit and suppression of truth.

CHAPTER ONE (8 MINUTE READD)

What's Agent Orange Got To Do With It?

The Earth is round, not flat, as everyone knows. And it rotates around the sun. Can we all agree on those two? Can we also agree that cigarette smoking causes cancer in some people? And gravity is the force that pulls objects toward the center of the planet?

There was a time when all these beliefs were hotly debated; today, they are all considered facts. Another thing everybody knows is that there is no link between childhood vaccines and autism. Everybody knows that notion has been debunked. Despite everybody being quite certain vaccines don't contribute to autism, since that possibility has been debunked, just about nobody can answer these two simple questions.

Who debunked it?

How was it debunked?

Most people can't answer even one of those questions, and that includes most pediatricians. Perhaps it was debunked so completely and obviously (known as being 'thoroughly debunked') that there's no need to ever wonder how or why we are all so sure. It would be like questioning gravity, except that the theory of gravity can be tested by any of us all day long. The idea that the vaccine-autism link has been debunked is more like an article of faith, akin to Adam and Eve being the first two people on Earth.

But this is not religion; this is science, right? And that raises a third question:

> Where did the idea that vaccines might be among the contributors to the increase in autism come from in the first place, such that it needed to be debunked?

For context, here are a few facts most people would agree aren't at all controversial, since they come directly from Federal public health agencies:

> In the 1950s, the rate of autism in American children was about ***one in 10,000***. In the late 1980s the rate took off, so that by the year 2000, ***one in 150*** children were diagnosed with autism. By 2023, the rate of autism in American children was ***one in 36***. (In California, it's currently ***one in every*** 22 children.) [1] [2]

Because these spiking numbers are so alarming, because this epidemic is so destructive to so many children and families, Federal public health authorities must have learned exactly what causes autism.

But they haven't. ("Scientists don't know exactly what causes autism spectrum disorder." — National Institutes of Health) [3]

Surely there must be effective drugs for treatment.

But there aren't. ("There are no medications that treat the core symptoms of ASD." — CDC) [4]

Surely there must be wide agreement on what exactly autism is.

But there isn't. There are no consistent biomarkers, no consistent physical characteristics, no blood or urine tests to confirm diagnoses — because autism is not a distinct disease. It's called a disorder, and was previously called a syndrome — a collection of observed symptoms subject to interpretation. One doctor might diagnose autism, and another doctor seeing the same patient might say it isn't autism, or say it's mild autism, or say it's nothing. Some children labeled and categorized with the diagnosis of autism appear to be fine, while others have dramatic and profound neurological disorders that require attentive care round-the-clock.

Though government scientists can't hold firm to a definition of autism and don't know what causes it, they claim with fist-clenching certainty to know what does *not* cause autism. How are they so sure? Because, as we're all aware, any link between vaccines and autism has been debunked. Since you probably don't know a single person who can say how it was debunked, or by whom, let's get that foundational question out of the way now.

> The possibility that any child's autism could be linked to any vaccine or combination of vaccines was debunked by the Institute of Medicine (IOM).

---

1  2  3  4 

What's the Institute of Medicine? Many people assume it's a prestigious government entity, but it's actually a totally private organization that hires members of various professions to examine issues pertaining to public health. Though IOM is often paid by government, it's also sometimes paid by private industry, including Pharma. As a private organization, IOM is not required to disclose how much it was paid for a given review, or who paid, or how much the hired experts were paid.

News media companies have tended to describe the Institute of Medicine as authoritative, independent, prestigious, respected, and the gold standard. IOM describes itself as unbiased, objective and evidence-based, and describes the experts it hires with words like *esteemed* and *renowned,* right up to *distinguished* and *eminent.* As you read on, it'll be your call to decide how much trust to invest in IOM.

Similarly, you'll decide how much trust to invest in me. On the one hand, I'm not a doctor, and on the other hand I'm not opining on medical science. I am a criminologist sharing information about corruption, crime and deceit, and I am a behavioral scientist sharing a few insights about human behavior and human nature. And I am a parent. I have followed my curiosity and skepticism, exactly as you can. In any event, the reader is not asked to trust me on anything, since citations with links to source material are provided throughout these pages.

Before we look at how and why IOM debunked the vaccine-autism link, let's take a quick look at another claim they debunked, the claim that the biological weapon Agent Orange caused sickness in some Vietnam Veterans, and birth defects in some of their children.

Agent Orange might seem irrelevant to the vaccine-autism issue, however IOM used some of the very same people and applied the very same process to reach the same conclusion for both Agent Orange and vaccines. And in both instances, their debunking conclusions were extensively marketed to the American public in the same way.

A quick bit of history:

For almost ten years, the US military sprayed Agent Orange onto jungles and people in Vietnam. An ingredient known as TCDD is the most toxic form of dioxin, and causes severe injury, including malformations in test animals, and (no surprise) in people too.[1] [2]

1  2 

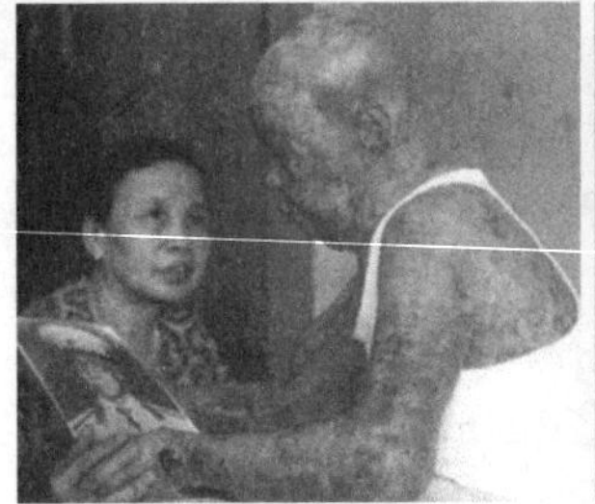

When the US was accused of violating the Geneva Protocol that regulated the use of chemical weapons, our government argued that Agent Orange was not a chemical weapon. They called it an herbicide useful for destroying food crops and jungles that afforded concealment for the enemy. It wasn't meant to hurt anyone, they insisted.

After studies showed that Agent Orange caused birth defects in test animals, the Department of Defense promised to reduce its use, though it would be years before it was officially suspended. Even the official suspension didn't end it, alas, because some military units falsified reports to conceal their continued use of Agent Orange. Eventually the military really did stop using it in Vietnam because, well, the war in Vietnam ended. By that point, many returning veterans had reported debilitating health problems in themselves and their children, affirmed by ever-expanding and terrible toxicologic reports about TCDD. [1]

To resist and reject these claims, the US Government wrote a check to IOM. Whether debunking the crazy idea that injecting mercury into children could possibly cause any neurological injury, or debunking the crazy idea that a military weapon containing dioxin could possibly be linked to health issues, IOM sang the same dreadful dirge. For example, they used the same approach in both scientific inquiries. ('Scientific inquiry' is a phrase that stretches those two words to the breaking point). They began both projects by setting forth what would *not* be included in their ostensibly deep explorations.

In their study of Agent Orange, IOM stated that they would not consider "toxicologic studies," because, you know, what possible bearing could toxicology have on the topic of toxicity? Similarly, in their study of vaccine safety, their report stated at the outset that they wouldn't "recommend a change in the licensure, scheduling, or administration of a vaccine." Meaning, their deep and comprehensive study

---

1 [QR code]

of childhood vaccines wouldn't propose anything about the administration of... childhood vaccines.

After one of its many Agent Orange reviews, the Institute of Medicine announced their bold conclusion: Studies into the reproductive history of individuals who'd been exposed to dioxin were [wait for it...] needed. Studies were needed. Their subsequent Agent Orange reports of *2006, 2008, 2010, 2012* and *2014* all reached that same dramatic conclusion: Studies needed.

Finally in 2018, the gloves came off when the IOM reaffirmed that studies were needed, but this time recommended something new: "**further specific study of the health of offspring of male Vietnam veterans**."

The bold font is theirs, by the way, because these folks wanted it clearly understood that when they conclude (five times) that further study is needed before they can conclude anything else, they really mean it. The closing words in their report didn't pull any punches, perhaps because they never landed any punches in the first place.

> "There are many questions regarding veterans' health that cannot be adequately answered by examining superficially analogous exposures and outcomes in other populations. It is only through **research on veterans themselves** that the totality of the military service experience can be properly accounted for."

Coleen Boyle

I find nothing to ridicule in that intelligent sentence above — other than that it took the vaunted Institute ***22 years*** to get there.

Though no parent would want the Government assessing childhood vaccines the same way it assessed Agent Orange, that's exactly what happened. And both projects involved two of the same central figures from the CDC: Dr. Frank DeStefano and Ms. Coleen Boyle, PhD.

Frank DeStefano

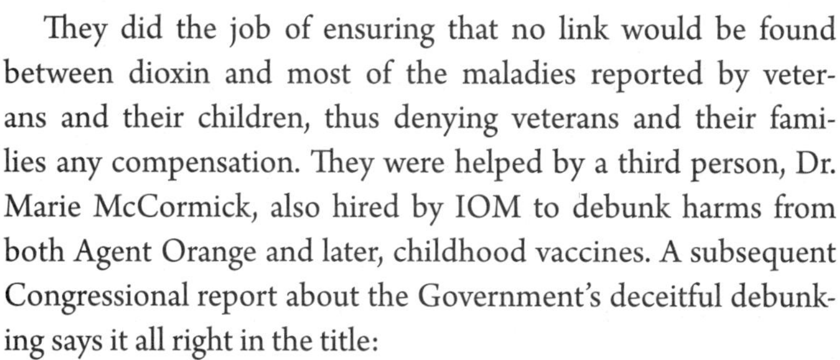

They did the job of ensuring that no link would be found between dioxin and most of the maladies reported by veterans and their children, thus denying veterans and their families any compensation. They were helped by a third person, Dr. Marie McCormick, also hired by IOM to debunk harms from both Agent Orange and later, childhood vaccines. A subsequent Congressional report about the Government's deceitful debunking says it all right in the title:

> "The Agent Orange Cover-up: ***A Case of Flawed Science and Political Manipulation.***"

Among its many findings:

- The Government "had secretly taken a legal position to resist demands to compensate victims of Agent Orange exposure..."
- CDC's work was "based on erroneous assumptions and a flawed analysis" [1]

Famed Navy Admiral Elmo R. Zumwalt Jr. led a review that was independent of the Institute of Medicine, and then testified before Congress:

> "The sad truth which emerges from my work is not only that there is credible evidence linking certain cancers and other illnesses with Agent Orange, but that government and industry officials credited with examining such linkage ***intentionally manipulated or withheld compelling information*** of the adverse health effects associated with exposure to the toxic contaminants contained in Agent Orange." [2 (pg 37)]

The Admiral's testimony was all the more powerful because it was deeply personal: He himself had ordered the use of Agent Orange in Vietnam — and his own son was among the soldiers who died from it.

For our purposes here, the Government's intentional manipulation and withholding of the truth starkly demonstrates that the Government has done such things. Readers who can embrace that idea will be open to the possibility that it might happen again when the same players at IOM use the same methods for the same money paid by the same client for the same reasons. All that differs is the toxin. [3] [4] [5]

After being roundly discredited for their shammy work on Agent Orange, were Dr. DeStefano and Ms. Boyle demoted by CDC? Nope. Were they fired? Not exactly. He was promoted — from Agent Orange to... **childhood vaccines.** [6]

After Dr. DeStefano was caught concealing study results that showed increased risk of autism from a vaccine product, he was promoted again — and not to just any position, but to Director of the optimistically-named Immunization Safety Office. (Recently, he was the perfect choice to find the obviously preferred answer to another pressing question: Might cardiac injury from mRNA vaccines be a serious problem?)

---

1  2  3  4  5  6 

And Ms. Boyle? Despite or perhaps because of her deceitful work on Agent Orange, she too was promoted at CDC, eventually to Director of the National Center on Birth Defects and Developmental Disabilities. Agent Orange prepared her well for her work on birth defects, causing as it does... birth defects. And childhood vaccines likely prepared her well for understanding developmental disabilities, causing as they too-often do... developmental disabilities.

Before blindly honoring the Institute of Medicine's debunking of any link to any ingredients in any vaccine products and any neurological injuries to any children, it would seem fair to consider if any more recent studies have debunked IOM's position. Wouldn't you know it, there's a whole bunch of published peer-reviewed science in that category, the most recent from 2025:

> A study of 47,155 nine-year-old children found the ***vaccinated children had far higher rates of autism*** than the unvaccinated (2.8 percent vs 1.1 percent), and that "vaccination was associated with ***significantly increased*** odds for all measured neurodevelopmental disorders," and that "vaccination is ***strongly associated*** with increased odds of neurodevelopmental disorders." [1]
>
> Another study looked at medical records from four HMOs, showing that infants exposed to greater than 25 µg of mercury in vaccines at the age of one month were 7.6 times more likely to have an autism diagnosis than those not exposed to any vaccine-derived organic mercury. [2]
>
> Review of 165 studies that focused on thimerosal, an ethylmercury compound in many childhood vaccines, "found it to be harmful." Sixteen of the studies specifically examined the effects of thimerosal on infants and children with reported outcomes of "developmental delay and neurodevelopmental disorders, including tics, speech delay, language delay, attention deficit disorder, and autism." [3]
>
> "Autism: A novel form of mercury poisoning" [4]
>
> Study found higher brain mercury levels after ethylmercury exposure [5]
>
> "Mercury is among the most harmful heavy metals to which humans can be exposed... the human body lacks effective mechanisms to excrete it." [6]

1  2  3  4  5  6 

> "Exposures to neurotoxicants such as lead, mercury, and pesticides can have a particularly detrimental impact on brain function and in turn lead to... learning and developmental disabilities."[1]

(See Appendix #5 for studies that compare the occurrence of neurodevelopmental disabilities in vaccinated children versus unvaccinated children.)

When there is a lot of research to undo, when there are so many forbidden facts to dissolve, you need something more than mere science. Luckily for the Government, IOM had a time-tested process — and not just from Agent Orange.

1

CHAPTER TWO (9 MINS)

Cancer from Baby Powder
Gulf War Syndrome
Silicone Implants
Gulf War Illness
Anthrax Vaccine
Burn Pits
SIDS

All Debunked!

In 2011, the Institute of Medicine was hired to debunk another reality the Government didn't like: serious health consequences from exposure to burn pits in Iraq and Afghanistan.

The IOM convened a committee to settle the issue once and for all. They set out first to compose five possible categories for describing the conclusion they'd soon reach:

- Sufficient evidence of a causal relationship
- Sufficient evidence of an association
- Limited/suggestive evidence of an association
- Inadequate/insufficient evidence to determine whether an association exists
- Limited/suggestive evidence of no association

You'll never guess which category they finally settled on. It was a mixed model, and just what the Government ordered: "Inadequate/insufficient evidence of an association between exposure to combustion products and cancer, respiratory disease, circulatory disease, neurologic disease, and adverse reproductive and developmental outcomes in the populations studied." [1] [2]

With a wave of their diplomas and an authoritative pronouncement, the IOM committee debunked the idea that breathing smoke from burn pits caused cancer or respiratory disease, or any disease at all. Heck, breathing that smoke might even have been healthy. (President Joe Biden nonetheless attributed the cancer death of his son, Beau, to burn pit exposure in Iraq. [3])

With this track record, it's no surprise that the Government hired the Institute of Medicine to debunk another crazy notion, this time the possibility that Gulf War Syndrome, which affected a huge percentage of Gulf War veterans, might have resulted from anything used by the US military.

One possible culprit was an ingredient in vaccines given to soldiers, as reported by *The Guardian*:

> "The common factor for the 275,000 British and US veterans who are ill appears to be a substance called squalene, allegedly used in injections to add to their potency. Such an action would have been illegal. Squalene is not licensed for use on either side of the Atlantic because of potential side effects." [4]

Before you consider whether squalene might be among the causes of Gulf War Syndrome, a Wikipedia article can save you time:

> "Attempts to link squalene to Gulf War Syndrome ***have been debunked***."

That settles that. What is the authoritative source for such a definitive pronouncement? The Institute of Medicine, of course. [5]

People with the patience to read IOM's report on Gulf War Syndrome won't see any mention of squalene until they've chewed through nearly 300 pages; it's in reference to a study of rats injected with a squalene compound:

> "Five of twenty animals immunized with this combination ***died within 1 to 5 days***... the animals suffered from shock and cardiovascular collapse."

That ugly result obviously had to be debunked before the sun went down, so IOM resurrected a favorite strategy for avoiding topics. Having first said that squalene was "an issue the committee was asked to address," they instead ran in the other direction:

> "The committee was not asked to draw conclusions on the strength of the evidence for an association between exposure to squalene and adverse health effects."

If not drawing conclusions on a possible association to adverse health effects, what in the world was this 400-page excursion all about? Debunking any possible association, of course.

On Page 308, there's another study of rats unlucky enough to be injected with squalene, this one showing that the injections triggered swelling of the brain that led to disturbance of the central nervous system. Pay no attention to the fact that disturbance of the central nervous system is among the symptoms of Gulf War Syndrome.

On Page 309, we learn that another adverse reaction was "so pronounced that researchers have coined the term Squalene-induced Arthritis." Pay no attention to the fact that arthritis is among the symptoms of Gulf War Syndrome.

On Page 311, we learn of a study that injected hundreds of people with two vaccine types, one containing squalene and one not. Myalgia was twice as common in those receiving the vaccine that contained squalene. Pay no attention to the fact that myalgia is among the symptoms of Gulf War Syndrome.

The report describes a study into the most relevant group, actual Gulf War veterans and military employees. Among those who developed chronic illness, squalene antibodies were found 95% of the time, and among those who *didn't* develop chronic illness, squalene antibodies were never found.

Results like that call for emergency debunkment, and IOM delivered it just a page later:

> "The committee does not regard this study as providing evidence that the investigators have successfully measured antibodies to squalene."

So there.

You've now had a brief look at the source material Wikipedia relied upon when pronouncing with such certainty that "Attempts to link squalene to Gulf War Syndrome have been debunked." Remember this the next time you read a confident debunking pronouncement in some media report — and certainly remember it whenever the original source is the Institute of Medicine. [1]

As long as there were sick veterans who needed to be reassured that they were probably just crazy, IOM kept up their tireless effort to find nothing. They published expensive reports on Gulf War Syndrome in 2000, 2003, 2004, 2005, 2006, 2007, 2008 and 2010, and all of them "did ***not find evidence*** that would support a confident attribution of the array of unexplained symptoms reported by veterans of the 1991 Gulf War to any specific chemical, biologic, or physical exposure." [2]

Year after year, IOM just couldn't find evidence to support linking anything to anything or anybody (or any body). If Gulf War Syndrome existed at all —and IOM never acknowledged that it did— the evidence they never found was sufficient to confidently debunk anything for which the US Government could be blamed.

IOM often succeeds at meeting the Government's wishes by outright excluding key areas from their inquiry. For example, even after years of supposedly studying Gulf War Syndrome, they stressed that

> "IOM Committees were NOT asked to determine whether a unique Gulf War syndrome exists, or to make judgments regarding the veterans' levels of exposure to putative agents." [3]

In other words, this was all busywork and distraction since they ***avoided the two biggest elements of the topic***.

The next debunking the Government undertook followed persistent complaints about the safety of the anthrax vaccine given to soldiers. Who'd the Pentagon bring in? The chairman of a past IOM committee, of course. Dr. Gerard Burrow offered his authoritative opinion in a four-page letter that concluded, "The anthrax vaccine appears to be safe, and offers the best available protection."

Though FDA described that vaccine's manufacturing conditions as "deficient," Dr. Burrows explained that deviations in product lines would not impact the quality or integrity of the vaccines. (Apparently, vaccine formulas don't have to be all that precise.) He based his confident proclamation that vaccines were safe upon the results of a study that wasn't yet completed, a study he promised "should be available in the near future."

---

1  2  3 

Later, when a Congressional committee pressed Dr. Burrows to explain more, he explained less:

> "Unfortunately, I do not believe I can make a significant contribution to the work of your committee... I was very clear that I had no expertise in Anthrax... the Defense Department was looking for someone to review the program in general and make suggestions, and I accepted out of patriotism."

Apparently, the Government was looking for a patriot, not an expert. [1]

During their decade of deep dives, IOM did find at least one interesting fact: Animal studies undertaken to test the anthrax vaccine "do not indicate whether the authors monitored for adverse consequences of vaccination." If you don't look for adverse consequences, you surely won't find them. And find them they did not. Unlike IOM, however, a Congressional Committee on Government Reform did look for adverse consequences, and found them, and ***soon after, mandatory anthrax vaccinations were halted***. [2]

Despite IOM debunking the idea that the military could possibly have caused any adverse consequences to any soldiers anytime anywhere, the Department of Veterans Affairs wasn't persuaded. They convened their own committee that sharply disagreed with IOM — and it didn't take them a decade to reach their conclusions.

> "Evidence strongly and consistently indicates that two Gulf War neurotoxic exposures are causally associated with Gulf War illness." [3]

Fortunately for science, the Institute of Medicine was able to quickly debunk that nonsense by explaining that even if a causal association was established, and even if evidence "strongly and consistently" indicated a link, "***the evidence was not robust enough***." [4]

Whew. Close one.

Drugs For the Healthy

Another possible culprit for Gulf War Illness was a drug called pyridostigmine bromide (PB). The US military ordered hundreds of thousands of soldiers to take PB pills for what they called "pretreatment." Turns out the pills purchased to

---

1  2  3  4 

protect troops from nerve damage ended up causing — wouldn't you know it — nerve damage. One significant study had a particularly unfortunate title, "Nerve Gas Antidote a Possible Cause of Gulf War Illness." [1]

The Department of Veterans Affairs concluded that the wonder-pills contributed to Gulf War Illness. So did the Rand Corporation. So did Boston University. So did Baylor University. But not IOM. As with the vaccine-autism link, IOM proclaimed that "the available evidence is not sufficient to establish a causative relationship."

And did the pre-treatment pills end up helping at all? Seems like No, in part because American soldiers never encountered the type of nerve gas the pills were designed for — and in larger part because the pills caused tremors, respiratory problems, cardiovascular problems, gastrointestinal problems, and neurological problems. [2]

PB pills remind us that medical interventions given to healthy people to prevent maladies that are very rare often fail the risk-benefit analysis. That lesson doesn't slow down commerce, apparently; at the end of 2024, Amneal Pharmaceuticals enthusiastically announced that their new "*PB-Extended Release*" had just received FDA approval for use by the military. Because of requirements from the Securities Exchange Commission, Amneal's breathless press release had to include this buzz-kill: Their wonder-drug for protecting solders from poison gas sometimes causes "breathing difficulties and loss of consciousness." What a coup. [3]

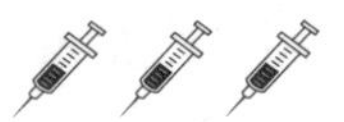

Another IOM committee was convened to debunk the notion that vaccines might be linked to Sudden Infant Death Syndrome (SIDS). And debunk it they did.

SIDS is defined as "the sudden death of an infant under 1-year of age, which remains unexplained after a thorough case investigation, including performance of a complete autopsy, examination of the death scene, and review of the clinical history."

In other words, SIDS is the umbrella term used when the cause of death is unknown and undiscoverable. It's a term, not an explanation — used when there

1  2  3 

is no explanation. It's a category of death, not a cause of death. A baby cannot be diagnosed with SIDS, and SIDS cannot kill a baby. Nonetheless, IOM set out to debunk the idea that any vaccine or combination of vaccines could possibly be the cause of these deaths for which the cause of death could not be determined. As with autism, they have no idea what caused these sudden deaths — but they sure know what didn't cause them:

> "The committee concludes that the evidence favors rejection of a causal relationship between exposure to multiple vaccines and SIDS." [1]

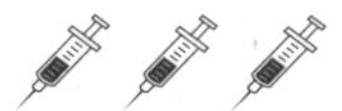

After more than 17,000 women and their families sued Dow Corning for harms from silicone implants, and after Dow Corning was required to pay billions in jury awards and settlements to more than 150,000 women — IOM came to the rescue. Though court cases had found that Dow Corning's products caused autoimmune diseases and neurological harms, [2] IOM's 1999 report debunked all that.

First, says IOM, there's no evidence at all:

> "**no evidence** linking implants to systemic diseases..."

Next, there's evidence — but it's flawed:

> "evidence that silicone breast implants cause neurologic signs, symptoms, or disease is lacking or flawed"

And finally, no matter what evidence there is or isn't, there's no basis for health concerns:

> "a review of the toxicology studies of silicones and other substances known to be in breast implants **does not provide a basis for health concerns**" [3]

Despite such a confident debunking by IOM in 1999, both the FDA and the World Health Organization have since recognized a cancer diagnosis associated with these breast implants. [4] [5]

The court cases that IOM debunked had all the usual suspects, including Dow Corning executives who were aware of the danger of silicone implants decades before the lawsuits started.[6] No matter, the debunking was just what the customer ordered: Prior to IOM's report, Dow Corning had been ordered to pay close to

1  2  3  4  5  6 

$4-billion in individual lawsuits, jury awards, settlements and class action suits. After IOM's report, not so much.

More recently, when it became important to debunk the silly notion that talcum baby powder could cause cancer, the assignment went to a company called Stantec ChemRisk instead of IOM. However, those hired experts were "guided by the US Institute of Medicine framework," an homage to IOM that's quite apparent when you read their conclusions:

> "Integrating all streams of evidence according to the IOM framework yielded classifications of *suggestive evidence of no association*... and *insufficient evidence to determine whether a causal association exists*..."[1]

Unfortunately for the folks who hired Stantec ChemRisk, the link between baby powder and cancer just wouldn't stay debunked. Worse for them, it's become *unbunked*. More on this later.

A pattern emerges: IOM is about messaging, not measurement or monitoring or microscopy or method. IOM helps the Government with crisis management, not science. When entrenched government programs and products are challenged, or when the truth would be politically inconvenient or too expensive, the Government (and big companies) have often hired IOM to debunk the problem away. No surprise in any of that.

What is surprising, however, is the extent to which these debunking scams have been effective, and the extent to which so many citizens and media companies are willing to parrot what authorities tell them. Once a corporate PR person or government official utters that magic word *Debunked*, no additional scrutiny is needed — or allowed. After that word is invoked, anyone curious or skeptical enough to keep asking questions is deemed stupid, crazy, a conspiracist — or all three.

With the curious and skeptical people neutralized, with debate prohibited, the truth is harder to find. And the stakes of finding the truth can be life or death, as public health officials and the medical establishment remind parents so frequently: *You don't want your child to get any of those deadly diseases that you could have prevented with vaccination.*

1

CHAPTER THREE (2 MINS)

Vaccines Are the Greatest Idea Ever Conceived

Most people would define a vaccine as *a benign and harmless injection that makes people immune to a disease, reliably preventing infection and transmission of diseases children are likely to be exposed to, and which, in the absence of vaccination, could be fatal or debilitating.*

That's the greatest idea ever. A safe injection (or series of injections) that doesn't harm anyone and provides immunity, preventing a child from getting and spreading a terrible disease — no sane person could oppose that idea. But does that definition actually apply to every vaccine marketed in the US? We know it can't apply to the vaccine products that were pulled from the market due to safety concerns (after millions of doses were administered to millions of people), but to which remaining vaccines does it apply?

Every medical product calls for a risk/benefit assessment — and it's always a matter of odds and probabilities. Like people investing in the stock market, parents have different feelings and ideas about taking chances. Unlike chances with money, however, this discussion is about the health and well-being of children. Just as different parents assess risks differently when their children go outdoors, swim, or play sports, parents also have different approaches to risks from disease, and risks from pharma products. Fill in the blanks to see what risk/benefit ratio works for you.

- I would give my child this particular vaccine if it prevents infection and transmission of the disease
  __ not at all
  __ very rarely
  __ most of the time
  __ all of the time

- and if the disease would otherwise cause death or life-altering health problems
  - ___ for zero healthy American children who get the disease
  - ___ for almost no healthy American children who get the disease
  - ___ for most healthy American children who get the disease
  - ___ for every healthy American child who gets the disease
- and if my child is likely to be exposed to this specific disease
  - ___ effectively never
  - ___ very rarely
  - ___ occasionally
  - ___ often
- **and only if the vaccine itself has caused a child's death or life-altering injury**
  - ___ no more than 5 times.
  - ___ no more than 25 times.
  - ___ no more than 100 times.
  - ___ no more than 1,000 times.
  - ___ no more than 10,000 times.

Once parents have selected the answers that match their comfort-level, the result might read something like this:

> I would give my child this particular vaccine if it prevents infection and transmission of the disease most of the time, and if the disease would otherwise cause death or life-altering health problems for most healthy American children who get the disease, and if my child is likely to be exposed to this specific disease often, and only if the vaccine itself has caused a child's death or life-altering injury ***no more** than 25 times*.

We'll revisit your selections a few chapters from now, and you can then decide which vaccines match your personal assessment of risk and benefit.

Many people believe vaccination is a simple choice, comparing a quick injection to a lethal disease. But that choice is not what's on the menu.

CHAPTER FOUR (12 MINS)

Forget About Autism

Before you forget about autism, consider this definition from a prominent researcher whose son has autism:

> Autism is a multifactorial syndrome; manifests primarily neurologically through varying degrees of social disengagement/lack of awareness; obsessive-compulsive traits; persistent and repetitive behaviors; language difficulties and deficits in physical coordination that interfere with the activities of daily living; symptoms from mild to extremely severe; majority experienced some obvious form of regression/loss of acquired skills during early childhood.

Even though all of that is true, I eventually had to abandon that definition. After you read the story of why, you might do the same.

Severe autism is, indeed, severe: children with that diagnosis can be completely nonverbal or restricted to merely repeating words and phrases, showing little to no interest in social engagement, and an inability to interpret basic social cues. People with severe autism are stuck in frequent and intense repetitive behaviors such as hand-flapping, rocking, or spinning, and have extreme difficulty with changes in routine or environment. Many have meltdowns and tendencies to self-harm requiring constant supervision. Many have hypersensitivity to sounds, lights, textures, touch and/or smells. A tendency to wander away from safe environments is common, often into traffic or bodies of water. (Drowning is among the top 2 or 3 causes of death for children with autism.) [1] [2] [3]

Not surprisingly, it's parents who give us the clearest picture:

> *"Our daughter is profoundly autistic. She is non-verbal, has major behavioral issues, is self-injurious... she cannot be left alone ever. Our daughter was a beautiful baby, who was developing normally, but who had obvious reactions to*

---

1  2  3 

her first two DPT vaccines. One left her leg swollen and red, and she developed a high fever and screamed after the other. But the doctors did not hesitate to give her a third DPT shot when she was 5 months old, and she went over the edge. She had the shot at 4:00 p.m., and by 6:00 p.m. she had a fever of 105 to 106 degrees... After that day, she was gone. Over the years, we have lost many friends and are distant from many family members because she is so hard to love and be around. It is very heartbreaking to see people reject her, and to have them suggest that we should have institutionalized her."

"My son (aged 44) has no speech, no functional use of his hands, and will no longer stand. He has a couple of seizures every day. His teeth had to be pulled because he would not allow anyone near his mouth to brush them. He is not potty trained. He is very sensory defensive, flaps his hands, and makes moaning noises."

"My daughter is a giant baby because although she is an overweight 18-year-old, she functions at the level of a 2-year-old. She has no life, really, compared to her peers. She has very little functional communication, and can only say a few words, like eat, or short phrases that she repeats incessantly. She is still in diapers, with no probability that she will ever be potty trained. My daughter now has frequent periods of frustration, extreme rage, and self-injurious behavior."[1]

While many children diagnosed with autism are profoundly disabled, others go to school, obtain college degrees, hold jobs, and have families. That massive pendular swing in severity —the spectrum— is part of why the word autism has come to have little consistent and concrete meaning.

When it was first identified (as Kanner Syndrome), autism was a collection of symptoms. Then it became a disorder —a functional impairment characterized by several different and differing symptoms and limitations— and then became a developmental disability, a term that overlooks the physical symptoms. Since there are no accepted biomarkers for autism, no blood or urine tests to confirm a diagnosis, no consistent physical features, it ends up being diagnosed by deduction rather than chemistry — by opinion — and of course, the opinions differ.

1

A Tale of Two Teenagers

The first teenager struggles to communicate and must frequently be restrained to prevent self-harm. Her parents insist that she wear a safety helmet due to her repetitive and forceful head-banging. She has frequent meltdowns, cannot control impulses, and cannot manage personal hygiene. She must be attentively supervised 24-hours a day.

The second teenager is shy and sometimes awkward around his peers, but he drives himself to high school each day, succeeds as a student, and will later establish a transnational tech company.

Though both teenagers are said to have autism, sharing the same diagnosis wouldn't make any sense if the word '*spectrum*' wasn't deployed as a catchall. Nobody would ask if vaccines cause shyness, or if environmental mercury is the reason some children are socially awkward — but once labeled with autism, both the mild and extreme ends of the spectrum are rolled into one artificial cohort.

The question that matters most doesn't require the word autism — and doesn't require much diagnosis either. It's a question we can all answer — because it's already been answered by government and Pharma: ***Do some vaccines and combinations of vaccines cause serious and debilitating neurological injury (brain damage) to some children?***

Conflicted stakeholders are able to slide around that very important question by using the words autism and spectrum to create a moving target that defies consensus or understanding. Nearly everything in life could be said to occur along a spectrum; one could attach the word to every possible sickness — but doing so would detract from precision and clarity. As it stands, autism is a bunch of symptoms and classifications and sub-categories with changing features, each of which is subject to someone else's interpretation and opinion.

In 2013, ostensibly to help clinicians better classify patients, three levels were added to the phrase Autism Spectrum Disorder (ASD).

Level 1: Requiring support
Level 2: Requiring substantial support
Level 3: Requiring very substantial support

"Requiring support" is a term that could be applied to every child in America at one time or another. But not every child has autism, or must be closely monitored round-the-clock lest he or she burst into a repetitive and violent self-harming cycle. Well-meaning and talented people have worked to bring greater accuracy and clarity to this confusing issue. Unfortunately, other stakeholders have worked to create, protect —and benefit from— diagnostic ambiguity, chaotic categorization, treatment complexities, and a labyrinthine, multifarious mess that even Wikipedia can't clean up — not even with 12,000 words and more than 400 citations. Consider this small sampling of the Humpty Dumpty phrases that all the King's horses and all the Wikipedia editors couldn't put together in a cohesive way:

- "diagnosis requires that symptoms cause significant impairment in multiple functional domains"
- "across multiple contexts"
- "a spectrum, meaning it manifests in various ways"
- "varying widely across different autistic people"
- "the spectrum is multi-dimensional"
- "not all dimensions have been identified as of 2024"
- "scientists argue that autism impairs functioning in many ways that are inherent to the disorder itself and unrelated to society"
- "disagreements persist about what should be part of the diagnosis"
- "there are meaningful subtypes or stages of autism"
- "there is no known cure for autism [and] some advocates dispute the need to find one"
- "a lack of high-quality evidence"
- "a set of closely related and overlapping diagnoses"
- "different categorization tools to define this spectrum"
- "change over time"
- "highly variable"
- "a broad and deep spectrum"
- "manifesting very differently from one person to another"
- "the spectrum model should not be understood as a continuum running from mild to severe, but instead means that autism can present very differently in each person"

After all that, Wikipedia shares a consensus among researchers; sounds promising until that consensus goes off the rails.

> "growing consensus among researchers that ***the established ASD criteria are ineffective descriptors of autism as a whole***"

Yes, we can see that. But on it goes...

- "deconstruct autism into at least two separate phenomena"
- "symptoms may not become fully manifest"

- "may vary according to social, educational, or other context"
- "range of diagnoses"
- "one diagnostic category for disorders that fall under the autism spectrum umbrella"
- "differentiates each person by dimensions of symptom severity, as well as by associated features"
- "separate social deficits and communication deficits into two domains"
- "presentation varies widely"
- "people who may not have ASD but do have autistic traits"
- "recent shift to acknowledge that autistic people may simply respond and behave differently than people without ASD"
- "conversational topics seem to be largely driven by an intense interest in specific topics"

This item about conversation applies only to some people with autism, given that CDC estimates 40% do not speak at all. [1]

Back to Wikipedia's valiant effort to clear things up:

- "two distinct groups with different social interaction styles"
- "ASD includes a wide variety of characteristics"
- "behavioral characteristics which widely range"
- "inability to identify biologically meaningful subgroups within the autistic population"
- "diagnosis of autism is based on a person's reported and directly observed behavior"
- "which criteria are used depends on the local healthcare system's regulations"
- "traits of Autism Spectrum Disorders often overlap with symptoms of other disorders"

Finally, Wikipedia throws in the towel:

> "The characteristics of Autism Spectrum Disorders make traditional diagnostic procedures difficult." [2]

Considering the list you just read, how in the world could IOM or anyone else reach definitive conclusions about this chameleon-like cacophony of categories? Nonetheless, they assure us there is no vaccine link. But no vaccine link to what?

---

1 
2 

To every autism symptom but only when they all appear together? To any autism symptom that appears alone? To a spectrum? The inquiry defies a viable scientific conclusion because the questions are not about science; they are about syntax.

Back to Wikipedia, as the editors take a stab at listing the various names for autism:

- Autism Spectrum Disorder
- autism spectrum condition
- formerly Kanner Syndrome
- autistic disorder
- childhood autism
- Asperger Syndrome
- childhood disintegrative disorder, and...
- pervasive developmental disorder not otherwise specified

How does Wikipedia say autism is diagnosed?

> "combination of clinical observation of behavior and development and comprehensive diagnostic testing completed by a team of qualified professionals including psychiatrists, clinical psychologists, neuropsychologists, pediatricians, and speech-language pathologists"

So that thing referred to in all those phrases above, that nebulous impenetrable imprecise word — that's the thing the Institute of Medicine ruled upon with such certainty. And depending upon the sentence it's placed into, that shape-shifting word can get a person branded as a conspiracy theorist or a fool or a science-denier. Conversely, a confident statement containing that perplexing word can crown someone as a highly intelligent, well-informed and responsible citizen who respects science and respects experts, and has a deep understanding of the whole issue. You're really smart if you say, "Oh that's been debunked by hundreds of studies," and you're slightly crazy if you say anything else about vaccines and autism. A lot is riding on how people use that word, that word they couldn't pick out of a line-up if their lives depended on it. And some people's lives do depend on it.

Since autism is an aggregation of symptoms, there is no meaningful distinction between the terms "autism" and "autism-like symptoms." But that nearly meaningless distinction ends up meaning everything to the beleaguered families who make the pilgrimage to Washington, DC in hopes of some mercy from the Government's Vaccine Injury Compensation Program.

Though refusing to even consider compensation in autism cases, the Government paid a settlement worth more than $100-million to the family of a (previously) healthy one-year-old baby girl who was injured by the MMR vaccine. Prior to vaccination, the baby had been walking and climbing; after the vaccination, it became clear she had a serious neurological injury and no clear path to recovery. [1]

What was the condition that the Government acknowledged was caused by the vaccine? A neurological disorder called encephalopathy that can be caused by exposure to toxins, including heavy metals like mercury, and can include swelling of the brain. [2] [3]

A team of researchers reviewed claims submitted to the National Vaccine Injury Compensation court for children who developed encephalopathy within days of receiving a measles vaccine, either alone or with other vaccines. Of 48 such children (age 10-months to 49-months), eight died. The remainder had mental regression and retardation, chronic seizures, motor and sensory deficits, and movement disorders. But it wasn't autism. Of course not. [4]

Similarly, the Government has acknowledged that vaccines can cause Residual Seizure Disorder.

One 9-month-old baby whose case was brought before the vaccine injury court had been rushed to the emergency room because of seizures shortly after vaccination. While the baby initially appeared to recover, the girl's mother soon noticed her daughter acting uncharacteristically remote, and then losing developmental abilities she had previously attained. Ultimately, this little girl developed epilepsy and developmental delays, serious enough that she requires constant care and supervision. After first refusing to compensate the family, and after a higher court chimed in, the vaccine injury court decided the girl's vaccine-induced seizures and developmental issues qualified for compensation. [5]

But it wasn't autism. Of course not. In another case, the vaccine injury court awarded compensation for a child's encephalopathy following a DTaP vaccine. A Federal Circuit Court found that the child had two diagnoses: medical encephalopathy ***and behavioral autism***. This is not the only time the vaccine injury court paid compensation for neurological harms when autism was also present. But it wasn't autism. The distinction is more about the words than the facts — but the words determine who gets help and who doesn't. That's because, unlike autism, the vaccine injuries of encephalopathy and Residual Seizure Disorder have been

---

1  2  3  4  5 

officially acknowledged by the Government. Other than that one difference, you might find the three conditions rather similar:

- All three conditions involve abnormal brain function.
- All three conditions manifest in early childhood.
- All three are strongly associated with developmental delay.
- Certain single-gene disorders are associated with both epileptic encephalopathy and autism.
- Some genetic mutations are associated with both epilepsy and autism.
- Epilepsy is common in children with autism (26%), while very uncommon among children in general (1%).
- All three conditions produce abnormal EEG patterns.
- All three conditions can have long-term effects on cognitive and adaptive functioning.
- All three can involve altered behavior patterns.
- All three can include Attention Deficit Hyperactivity Disorder.
- All three can adversely effect language and social communication.
- All three are strongly associated with regression and loss of acquired skills.
- All three conditions frequently involve seizures.
- 70% of children with severe seizures (infantile spasms) develop autism/intellectual disability. All three conditions can co-exist, with epilepsy being common among patients with both autism and encephalopathy. [1] [2] [3] [4] [5] [6] [7] [8]

Encephalitis (inflammation of the brain) is another condition the vaccine injury court has acknowledged can be caused by some vaccines.[9] By this point, you'll likely be able to predict the symptoms associated with encephalitis:

- permanent mental impairments
- changes in nature
- cognitive impairments
- motor function disorders
- neuropsychiatric issues
- personality and behavioral changes

- recurrent seizures or epilepsy
- neurological complications
- speech and language problems
- ***intellectual disability, especially in children exposed at a young age***

But of course it's not autism; we know it can't be autism because of the last symptom on the list that describes encephalitis:

"autism-like behavioral problems" [1]

Since autism is defined almost entirely by "autism-like behavioral problems," perhaps it's time to acknowledge that encephalitis, encephalopathy, Residual Seizure Disorder and autism can all present with the same symptoms, can all be associated with inflammation of the brain, and inflammation of the brain and neurological side-effects have been acknowledged by the vaccine injury court as being caused by vaccines in some children.

A quick note on inflammation: ***The very purpose of vaccines is to induce an inflammatory response.*** [2] [*]

An excellent 2011 article in the Pace Environmental Law Review put it plainly:

> "About half of the eighty-three reviewed cases have encephalopathy, residual seizure disorder, and autism. The other half of the reviewed cases have residual seizure disorder and autism. ***There is no obvious distinction in symptoms or gravity of injury among these cases.***"

The vaccine injury court acknowledged autism or autism-like symptoms associated with vaccine-induced encephalopathy and seizure disorder in at least 21 of its rulings. [4]

> "Based on this preliminary assessment, there may be no meaningful distinction between the cases of encephalopathy and residual seizure disorder that the [court] compensated over the last twenty years, and the cases of 'autism' that the [court] has denied. If true, this would be a profound

* "The expression of pro-inflammatory molecules is not only restricted to muscle... inflammation in the circulation can affect other body systems to cause systemic side-effects... Balancing the beneficial versus the detrimental effects of these inflammatory events is necessary to keep reactogenicity at clinically acceptable levels. [3]

1  2  3  4 

> injustice to those denied recovery and to all who have invested trust in this system that Congress created."

Given these realities, might it make sense to stop carving out collections of relevant symptoms simply because those symptoms were at some point classified under the word autism? Remember: same symptoms, same possible cause (inflammation of the brain or reaction to a toxin), and the Government acknowledges that these same symptoms and causes are sometimes associated with some vaccines given to some children — **so long as the neurological disorder is not named autism**.

The Pace article uses a phrase that's simple and clear: vaccine-induced brain damage.

When we set aside the word autism (even just momentarily), we can speak instead of neurological and neurodevelopmental disorders and their causes. We can speak of brain damage, without regard to the labels. Any other approach distorts, minimizes and denies the severe conditions being lived by many thousands of children (many of whom are now adults) and their families. If we set aside that one word —autism— it's instantly easy to answer the question of whether some vaccines sometimes cause serious neurological injury, brain damage and persistent neurodevelopmental disorders in some children.

In fact, the vaccine-makers' own package inserts have already answered that question many times over the years, frequently having had to acknowledge the neurological side effects from their products, but never ever using the word autism — not even when the harms involve inflammation of the brain:

MMR Vaccine

- Seizures
- Encephalomyelitis (inflammation of the brain)
- Transverse myelitis (inflammation of the brain and spinal cord)
- Syncope (loss of consciousness)
- Polyneuropathy
- Ataxia (lack of voluntary muscle coordination, speech changes, abnormal eye movements — all indicating brain dysfunction)
- Guillain-Barré Syndrome
- Progressive Neurological Disorder

Varicella Vaccine (chickenpox)

- Ataxia
- Encephalitis
- Transverse myelitis

Guillain-Barré syndrome
Seizures
Bell's Palsy
Stroke
Meningitis

Hepatitis B Vaccine
Multiple sclerosis (neurological disease associated with inflammation of the brain)

Influenza Vaccine
Guillain-Barré syndrome
Seizures

Pneumococcal Conjugate Vaccine
Seizures

Oral Polio Vaccine
Vaccine-associated paralytic poliomyelitis

Meningococcal Vaccine
Guillain-Barré syndrome

HPV Vaccine
Guillain-Barré syndrome

DTP/DTaP Vaccine (Note: DTaP currently used)
Seizures
Prolonged convulsions
Encephalopathy
Neuropathy
Guillain-Barré syndrome
Hypotonic-hyporesponsive episodes (linked to developmental delays)
Lowered consciousness
Persistent neurologic symptoms
Unresponsiveness
Coma
Progressive neurologic disorders

Covid-19 Vaccines
Hemorrhagic Stroke
Guillain-Barré syndrome
Transverse myelitis

Encephalitis
Meningitis
Bell's Palsy
Seizures
Convulsive Disorders [1] [2] [3] [4] [5] [6]

(Since all the vaccines listed above are recommended for infants and children in America, it's worth noting a 2025 peer-reviewed study about the deaths of three children within 24-hours of receiving routine childhood immunizations. The authors call for reevaluation of the "risks and benefits of currently approved vaccines" and a review of the childhood vaccination schedule.) [7]

Important: The list above is not an account of all side-effects — far from it. Rather, it's a list of only those side-effects that are neurological, many of which can have crossover with autism. Since vaccine-makers already acknowledge these specific neurological injuries, why is the entire discussion of autism distorted and sabotaged by that one unscientific, imprecise, misunderstood and often misused word?

---

1 2 3 4 5 6 7

> Pediatrics. 1998 Mar;101(3 Pt 1):383-7. doi: 10.1542/peds.101.3.383.

Acute encephalopathy followed by permanent brain injury or death associated with further attenuated measles vaccines: a review of claims submitted to the National Vaccine Injury Compensation Program

R E Weibel [1], V Caserta, D E Benor, G Evans

Objective: To determine if there is evidence for a causal relationship between acute encephalopathy [sudden and severe change in brain function] followed by permanent brain injury or death associated with the administration of further attenuated measles vaccines, mumps vaccine, or rubella vaccines, combined measles and rubella vaccine, or combined measles, mumps, and rubella vaccine, the lead author reviewed claims submitted to the National Vaccine Injury Compensation Program.

Results: A total of 48 children, ages 10 to 49 months, met the inclusion criteria after receiving measles vaccine, alone or in combination. **Eight children died, and the remainder had mental regression and retardation, chronic seizures, motor and sensory deficits, and movement disorders**. The onset of neurologic signs or symptoms occurred with a **nonrandom, statistically significant distribution of cases on days 8 and 9**.

Conclusions: ***This clustering suggests that a causal relationship between measles vaccine and encephalopathy may exist as a rare complication of measles immunization.*** [1]

---

1 [QR code]

CHAPTER FIVE (2 MINS)

Deceptive Duplicitous Distorted Double-Dealing Definitions

Having just devoted an entire chapter to the tortured definition of the word *autism*, here's another key word that's been defined and redefined (and tortured) by public health officials:

CDC's ORIGINAL DEFINITION OF VACCINE

"a product that stimulates a person's immune system to produce immunity to a specific disease, protecting the person from that disease"

CDC's NEW & IMPROVED DEFINITION OF VACCINE

"a preparation that is used to stimulate the body's immune response against diseases"

What happened to a vaccine ***producing immunity to the disease***? What happened to ***protecting the person from getting the disease***? Those two ideas died of Covid in 2021.

Since the whole world could see that Covid vaccines didn't provide immunity and didn't protect people from getting the disease, the solution was to have everyone start pretending that vaccines were never intended to make people immune to a disease. [1]

Here's another word that had a useful makeover:

1

PANDEMIC (defined by the World Health Organization)

"An influenza pandemic occurs when a new influenza virus appears against which the human population has no immunity, resulting in ***several simultaneous epidemics worldwide*** with ***enormous numbers of deaths and illness***."

NEW & IMPROVED DEFINITION OF PANDEMIC (WHO)

"An influenza pandemic may occur when a new influenza virus appears against which the human population has no immunity."

The virus doesn't have to be widespread, and it doesn't have to be deadly or make even a single person ill. The WHO may now declare a pandemic and acquire all the associated attention, control and fun without outbreaks around the world, without anyone being ill and without even a single death. [1]

But the public, of course, understands the word *pandemic* to mean lethal pathogens are spreading worldwide causing severe illness and death — an emergency. And the public understands the word *vaccine* to mean a drug that gives you immunity from a disease and protects you from that disease.

We can be sure our clever public health officials and their Pharma Phriends are fully aware that changing the technical definition doesn't change what's conjured in the public mind when these words are invoked. Since there is no shared definition of *vaccine,* and since the word *autism* has also been destroyed, then no amount of evidence and no argument or experience or study or opinion can ever gain any traction. And that, apparently, is exactly the way Pharma and public health officials like it — because they created it.

1

CHAPTER SIX (6 MINS)

Brain Damage by Any Other Name

As a thought experiment, imagine that people just stopped arguing, that everyone instantly agreed that vaccines don't contribute to autism. Imagine everyone accepted that all possible links between vaccines and autism had been persuasively debunked by an unconflicted IOM, and imagine everyone came to respect IOM's sage process. Imagine that those thousands of parents who saw their children begin regressing soon after vaccination suddenly realized they were mistaken about what happened.

Not even that universal agreement about autism would change the established fact that ***vaccination is sometimes linked to serious neurological injury*** (brain damage). Since our government and Pharma have already acknowledged this reality, all that's left to debate and discover is how many children are injured by vaccines, and which children, and which vaccines — and then try to find ways to prevent the injuries from continuing. In the world we're imagining, our Government's vaccine injury compensation programs would work to regain some of the confidence and trust they have lost. Here's part of how they lost it:

> When people in Japan claim injury or death from the Covid vaccines, the government there pays compensation in 87% of the cases. [1]
>
> In Thailand, the government acknowledges Covid vaccine injury and pays compensation in 82% of the cases.
>
> In Canada, the government acknowledges Covid vaccine injury and pays compensation in 22% of the cases.

---

1 [QR code]

> In France, the government acknowledges Covid vaccine injury and pays compensation for vaccine injuries in 17% of the cases. [1]
>
> But when people in the United States claim injury or death from those same Covid vaccines, our government agrees to compensation in **fewer than 1%** of the cases. [2]

For many decades, the US Government's resistance to acknowledging vaccine injuries has been strident and intense, with millions of dollars invested into debunking, attacking, denying and refusing. Even as the Government reluctantly pays compensation for cases of autism, or autism-like, or autism-adjacent, or indistinguishable from autism — they always deny it's autism.

> From: Bowman, David (HRSA) [mailto:DBowlan@hrsa.gov]
> Sent: Friday, February 20, 2009 5:22 PM
> To: 'dkirby@nyc.rr.com'
> Subject: HRSA Statement
>
> David,
>
> In response to your most recent inquiry, HRSA has the following statement:
>
> The government has never compensated, nor has it ever been ordered to compensate, any case based on a determination that autism was actually caused by vaccines. We have compensated cases in which children exhibited an encephalopathy, or general brain disease. Encephalopathy may be accompanied by a medical progression of an array of symptoms including autistic behavior, autism, or seizures.
>
> *Some children who have been compensated for vaccine injuries have shown signs of autism before the decision to compensate, or may ultimately end up with autism or autistic symptoms, but we do not track cases on this basis.*
>
> Regards,
>
> David Bowman
> Office of Communications
> Health Resources and Services Administration
> 301-443-3376[113]

1  2 

The vaccine injury court publishes the National Childhood Vaccine Injury Act Reporting and Compensation Tables (VIT), a macabre document that lists the injuries associated with each vaccine product, and sets the precise time period during which the adverse event must commence in order to "qualify" for compensation. For example, when the Government awarded a family close to $2 million in damages, along with $250,000 a year to cover medical expenses throughout their injured daughter's life, the family's claim was able to succeed because the baby's seizures started 70 hours after a pertussis vaccination. If the seizures had started 73 hours after the injection, the court would have said, *Please take your lawyer and your daughter, and go home.* [1]

Since the Government has paid out billions of dollars to compensate for vaccine injuries and deaths, and since it makes payments for conditions that share so many features with autism, what is the official position of the court when it comes to autism claims?

Well, the National Vaccine Injury Compensation court, the entity created by Congress to consider autism claims… **currently refuses to hear any autism claims.** Why? We can logically speculate that it has something to do with the fact that there were once 5,400 autism cases on the waiting list, and if the court had ever endorsed the heretical notion that any vaccine or combination of vaccines might contribute to autism in any child, that decision would soon enough cost *hundreds of billions of dollars* in awards. The lifetime cost of care for a person diagnosed with severe autism is $3-million, on the low end. Since the rate of these neurological and developmental disorders is sharply increasing, whatever is causing it simply ***must not and cannot be vaccines***. If even just 10,000 such cases were compensated (and remember, there were already more than 5,000 on the waiting list twenty years ago, when the rate of autism was much lower), the cost would be $30-billion.

> *Since our government and Pharma have already acknowledged the reality, all that's left to debate and discover is how many children are injured by vaccines, and which children, and which vaccines — and then try to find ways to prevent the injuries from continuing.*

1

Conversely, if the vaccine injury court refuses to hear any autism claims, the cost is nothing. Problem solved.

There's also the matter of human nature, and the inhuman nature of powerful institutions. The Government and the healthcare industry and universities and prominent doctors have staked out and promoted the ironclad position that vaccines never cause widespread harms. It appears they will defend that position no matter what, even when lives are on the line. Of course, it's not their lives on that line.

So far, no amount of reality has been able to crack open why the Government insists that every single vaccine and vaccine ingredient is worth the risks — and plenty of people have tried. In 2004, Congressman Dan Burton convened hearings of the Congressional Subcommittee on Human Rights and Wellness.

Chairman Dan Burton:

> "Every time I talk to people who appear before the committee, either privately or in public forum, I say to them, would you mind if we just took the thimerosal, the mercury, and injected it into you like they did our kids? And they will say to you, 'Well, I don't think I want mercury injected into our bodies.' And these are doctors who say there's no harm being done. But they don't want mercury stuck in their bodies with a needle. Yet we do it to our kids every single day, and we do it to adults ... we find out that mercury is in the environment and they're saying, 'we have got to get it out of the environment because of the problems with the neurology of our population.' Yet we continue to put it into our bodies with needles."

During his committee hearings, Chairman Burton called many witnesses, including William Egan, at that time the Acting Director of FDA's Office of Vaccines Research and Review.

> Chairman Burton: Has thimerosal ever really been tested? Has thimerosal ever been tested by our health agencies?
>
> Mr. Egan: Only in those early tests that you know of that were done by Lilly.
>
> Mr. Burton: That was done in 1929. Let's follow up on that. In 1929, they tested this on 27 people that were dying of meningitis. All of those people died of meningitis, so they said there was no correlation between their death and the mercury in the vaccines. That is the only test that's ever been done on thimerosal that I know of. Can you think of any other?

> Mr. Egan: ***No, in people, no***. Except for accidental exposures over time.
>
> Mr. Burton: So we have mercury that's being put into people's bodies in the form of this preservative, and has been since the 1930s, and it's never been tested by our health agencies.

By 2002, thousands of parents had filed lawsuits claiming their child's autism came from thimerosal-containing vaccines. Some factions in Congress blocked those legal claims by surreptitiously inserting what was called the "Eli Lilly Rider" into the massive Homeland Security bill. The Rider wasn't to protect the homeland; it was to protect the makers of thimerosal.

Congressman Dan Burton was plenty shocked when he learned that the Eli Lilly Rider ordered the Justice Department, *in the supposed interest of national security,* to seal all the vaccine-related papers and documents (including secret transcripts you'll read about later).[1,2]

Bill Frist

The Rider, which galvanized parents nationwide into activism on behalf of their vaccine-injured children, barred any judge from issuing compensation to damaged children, and transferred all thimerosal cases to the vaccine injury court. A strong supporter of this outcome was Senator Bill Frist, who had introduced a nearly identical bill the year before. Eli Lilly had been a good friend to Senator Frist, not just by donating hundreds of thousands of dollars, but also being pal enough to purchase 5,000 copies of his book on bioterrorism. (Frist received thousands more from pharmaceutical companies immediately after the Homeland Security rider passed.)[3]

Safety Information

Hazard pictogram	  
Hazard statement	H300/310/330 - Fatal if swallowed, in contact with skin or if inhaled. H373 - May cause damage to organs through prolonged or repeated exposure.

Soon after leaving the senate, Frist joined the board of the Kaiser Family Foundation, later spent 10 years on the board of the Robert Wood Johnson Foundation (think Johnson & Johnson), and in more recent times

1  2  3 

he's been on the boards of a bunch of other health-industry corporations. Being the only medical doctor in the Senate at that time, Dr. Frist presumably knew all about risky pharmaceutical ingredients, including the one that comes with the warning, "Not for medicinal use," the same one known to be "toxic if swallowed, fatal in contact with skin, fatal if inhaled" — but okay to inject into babies. [1]

1

CHAPTER SEVEN (12 MINS)

RFK Jr & His Crazy Unhinged Questions

There are people who believe that the issue of a vaccine-autism link was invented by Robert F. Kennedy Jr, as if he awoke one morning inspired to ruin his career by asking whether mercury in childhood vaccines could cause brain injury. But that's not what happened.

Long before Kennedy gave the topic any attention, and long after he'd had all seven of his own children vaccinated, and long before Kennedy ever said a public word about vaccines, and long before he wrote his first book about mercury in vaccines, and long before public advocacy on behalf of children became his main mission, there were already many peer-reviewed studies and published scientific papers suggesting a link between mercury and neurological/ developmental disorders, including autism.

How did Kennedy end up grabbing onto Pharma's third rail? It started when he was traveling around the country giving speeches about the dangers of mercury in fish. As Kennedy explained to Joe Rogan, "These women start showing up at every lecture that I give, and they believe that their children had been injured by vaccines. They would say to me, '*If you're really interested in mercury exposure to children, you need to look at vaccines*.' Now, this was something I didn't want to do."

To get his attention, one of the mothers found his home in Hyannis Port, knocked on the door and said, "I'm not leaving till you read this." She placed a tall stack of scientific studies on the step.

"When I read that," Kennedy explains, "then I was like, okay, I got it... drop everything and do something about this."

Back when Kennedy worked on getting mercury out of fish, nobody ever said he was anti-fish. But as soon as he was interested in getting mercury out of vaccines, he was called an anti-vaxxer — a pejorative. He soon learned that if you ask questions about vaccines, you are going to lose a few things in short order:

- All opportunities to appear on TV (Pharma spends $22 **b**illion on media advertising each year [1])
- All opportunities to write articles and opinion pieces for major publications
- Friendships
- Relationships with your siblings (Really? Yes)
- Income
- Reputation
- Career

For nearly 20-years without pause —and up to the present day— Kennedy has been the target of what political operative Jay Carson calls THE PLAYBOOK. Carson, an advisor to President Clinton, Senator Clinton, Senator Daschle, Governor Dean, and Mayor Bloomberg, explains:

> *Big corporations hire people like me to implement the playbook. And here's the way the playbook works: First, they attack you broadly and they question your facts. They say you are lying, and it's ferocious, but if you keep on moving after that they switch to character assassination. They take on who you are as a person. They dig up everything bad in your past and leak it to the press. If you had a fender-bender, you're a reckless driver. If you paid a bill late, you're a deadbeat. And so on, and so on.*
>
> *Every part of your life goes under their microscope. They try to embarrass you. They try to make you say this fight isn't worth what its costing me, and you quit.*
>
> *If all that doesn't stop you — and it stops most people, but it didn't stop Bobby — they say you are a liar. If liar doesn't work, they say you are an antisemite and racist.*
>
> *Crazy or kook or crank or nutjob are their mainstays. That's their nuclear option. Because if they can get everyone to dismiss you as a wacko nutjob, everything you say is suspect.*
>
> *If all the publications and TV shows you trust tell you that someone is a crazy person, they must be a crazy person, right?*
>
> *Standing up to them exacts a real cost, and that's why they do it.* [2]

---

1 2

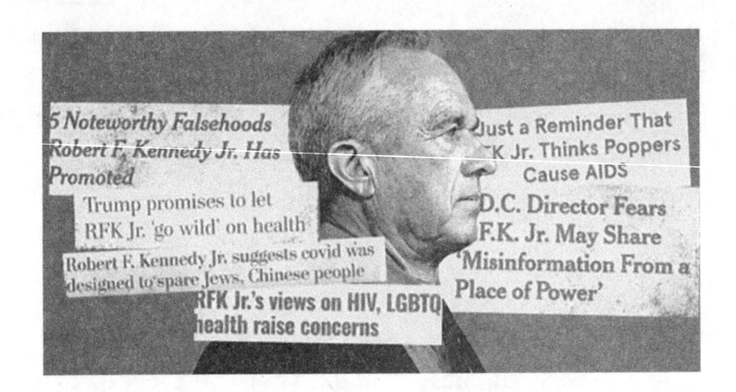

POLITICS

The Story About a Worm Eating RFK Jr.'s Brain Is Not As Funny (?) As It Sounds

PBS NEWS

Did a tapeworm really eat part of Robert F. Kennedy Jr.'s brain?

RFK Jr.'s vaccine quackery has no place in modern America

INTERNATIONAL · UNITED STATES

Anti-vax conspiracy theorist and JFK's nephew: Who is Robert F. Kennedy Jr.?

Elon Musk hosted RFK Jr. in a bizarre Twitter Spaces conversation littered with falsehoods and conspiracy theories.

POLITICS

RFK Jr. said doctors found a dead worm in his head after it ate part of his brain

International

ABC News Liberally Censors RFK Jr. Interview Over Vaccine Comments

18:17 GMT 28.04.2023 (Updated: 05:40 GMT 30.04.2023)

The next time you see stories like these ... you'll know it's just THE PLAYBOOK in action.

After a lifetime of being widely praised for his environmental activism, RFK Jr became a media-created pariah, the target of a massive, well-funded publicity campaign to sell one idea: Robert Kennedy Jr is not just wrong — he's crazy. When you can't knock down the facts, you knock down the people who repeat those facts. If someone's credibility and reputation can be destroyed, then few people will consider the questions he asks, or the answers he finds.

The tricky triad of Pharma, government, and corporate media have succeeded at morphing any discussions of vaccine safety into a discussion about Robert Kennedy Jr. But many other people have had concerns about the vaccine-autism link, including for example, uh oh, the Director of the National Institutes of Health (and Anthony Fauci's boss).

Bernadine Healy

Dr. Bernadine Healy, MD: *This is the time when we do have the opportunity to understand whether or not there are susceptible children, perhaps genetically — perhaps they have a metabolic issue, a mitochondrial disorder, immunological issue — that makes them more susceptible to vaccines plural, or to one particular vaccine, or to a component of vaccine, like mercury. The fact that there is concern that you don't want to know that susceptible group is a real disappointment to me. If you know that susceptible group, you can save those children. The reason why they didn't want to look for those susceptibility groups was because* ***they're afraid that if they found them, however big or small they were, that that would scare the public away.***

CBS News: It sounds like you don't think that the hypothesis of a link between vaccines and autism is completely irrational.

Dr. Healy: *So when I first heard about it, I thought well that doesn't make sense to me. The more you delve into it, if you look at the basic science, if you look at the research that's been done in animals, if you also look at some of these individual cases, and if you look at the evidence that there is no link,* ***what I come away with is the question has not been answered****.*

CBS News: Do you feel the Government was too quick to dismiss out of hand that there was a possibility of a link between vaccines and autism?

Dr. Healy: *I think the Government or certain public health officials in the Government have been too quick to dismiss the concerns of these families without studying the population that got sick. I haven't seen studies that focus on 300 kids who got autistic symptoms within a period of a few weeks of a vaccine.* [1]

Though Federal public health officials and vaccine manufacturers often say things like, "Millions of people have taken vaccines, and they are just fine," Dr. Healy's point is key: *Study the ones who are not fine.*

By way of comparison, most of us consider commercial air travel to be a reasonable risk. But: When a commercial airliner crashes, it is a big deal —a big deal— and every crash is subject to a comprehensive part-by-part study that often produces changes and improvements. All commercial jetliners are equipped with two black-box recorders that provide investigators with hundreds of flight characteristics used to reconstruct what happened.

Imagine the airlines took Pharma's approach: A flight from NY to London goes down in the Atlantic just before reaching Heathrow Airport. Imagine the airlines told the bereaved families that flight was 95% safe and effective, or "Most of our flights make it the whole way without crashing."

We all know there are risks with commercial air travel, and the ratio works for most of us. The fatality rate for air travel is currently around 17 deaths per billion passengers a year. (For comparison, a conservative rate for Covid mRNA vaccines is around *40,000* deaths per billion doses — yet no comprehensive study or action.)[2]

Unlike in commercial air travel, the Government has declined to do an honest study of individual or mass vaccine injuries and deaths — and they deny that most even occur. You can't hide an airline crash, but you can hide a rate of injury that's distributed across hundreds of millions of people.

---

1  2 

Years ago, when everybody but the Government became concerned about the spikey increase in autism (coincidentally concurrent with the spikey increase in mass vaccination), more and more parents put the two issues together. Thousands of parents directly observed and then reported that their young children experienced dramatic developmental decline immediately after being vaccinated, often by one vaccine in particular: Measles-Mumps-Rubella, known as MMR. And sometimes by combinations of vaccines given at the same time.

Too many reports from doctors, too many published papers, and too many damn parents with their damn experiences eventually meant that the Government had to do something. You might think the *something* would be finding out what causes autism — but it was quite the opposite. The Government decided to find out what *doesn't* cause autism. And no better source for discovering what doesn't cause something than the Institute of Medicine, which started chasing the reality of autism around the track until reality collapsed.

As with Agent Orange, burn pits, the anthrax vaccine, Gulf War Syndrome, Sudden Infant Death Syndrome, and breast implants, the Government was prepared in advance to make sure IOM's debunking would instantly become a giant news story. Soon enough, everybody knew the vaccine-autism link was debunked. After that, only fools or irrational people would question the esteemed experts — meaning there were very few questions, and none from the news media.

PUBLIC HEALTH

Report: Vaccines Are Safe, Hazards Few And Far Between

AUGUST 25, 2011 · 1:45 PM ET

PBS FRONTLINE

Report: No Link Between Vaccines and Autism

AUGUST 26, 2011

The National Academy of Sciences released a report yesterday concluding that "few health problems are caused by or clearly associated with vaccines."

HEALTH NEWS AUG. 26, 2011 / 8:37 PM

Study: No vaccine and autism link found

A committee of experts convened by the Institute of Medicine, a non-profit group outside the framework of the U.S. federal government to provide independent guidance and analysis to improve health, said few health problems are caused by or clearly associated with vaccines.

AUGUST 25, 2011 | 3 MIN READ

Childhood Vaccines Cleared of Autism, Diabetes Link in New Report

U.S. Institute of Medicine finds "very little evidence" of serious harm

Nature Magazine declared, "Vaccines are largely safe, and do not cause autism."

Their source? The Institute of Medicine.

In fact, IOM was the source for everybody who parroted that message — and it still is.

Promoting their work on vaccine safety, an IOM spokesperson said, "We looked very hard and found ***very little evidence*** of serious adverse harms from vaccines. The message I would want parents to have is one of reassurance."

Since that's the same "very little evidence" the Government found with Agent Orange, burn pits, the anthrax vaccine, Sudden Infant Death Syndrome, breast implants, and Gulf War Syndrome, I'm not sure how reassuring it ought to be to parents.

Let's get some clarity from the most prominent vaccine-advocate, Dr. Paul Offit, who has worked closely with Federal public health agencies for many years. Commenting on the IOM's conclusion, he said:

> "For those parents who are on the fence, this will be another piece of reassuring evidence, although I don't know how many more pieces of reassuring evidence you need."

This sentiment might have been more effective if Dr. Offit hadn't added a moment later that he was "uncomfortable as a scientist" with the Committee's methodology. "They're looking at case reports and trying to decide whether they think the evidence supports a link. That's an unusual way to do science, because now you're making it more subjective."[1]

> *"You can never really say 'MMR doesn't cause autism.' But frankly, when you get in front of the media, you better get used to saying it, because otherwise people hear a door being left open when a door shouldn't be left open."*[2]
>
> — Dr. Paul Offit, FDA & CDC Advisor

To IOM's credit, their 2011 IOM report on vaccine safety finally acknowledged a significant adverse effect from the MMR vaccine: brain inflammation. But autism? Impossible.

But is it impossible? In 2023, scientists from the University of Maryland School of Medicine confirmed that inflammation alters the development of vulnerable brain cells, and "***could explain how inflammation contributes to conditions like autism spectrum disorders.***"[3]

A series of studies in Italy demonstrated that brain inflammation is a key component of autism, and that "all vaccines can cause neuroinflammation." In fact, vaccines are designed to elicit an inflammatory immune response, so it's not surprising that vaccines cause brain inflammation in some babies.[4]

Aluminum, an ingredient in childhood vaccines, ends up redistributed to numerous organs, including the brain where it accumulates, and each new vaccine contributes more to this accumulation. In 2015, researchers studying this basic mechanism called it "a major health challenge, since it could help to define susceptibility factors to develop chronic neurotoxic damage."[5]

Many studies confirm that aluminum (like mercury) leads to chronic brain inflammation and neurotoxicity in some children.[6][7]

So... ***vaccines can cause brain inflammation, and brain inflammation can cause autism***, but you'd have to be as crazy as RFK Jr to even wonder if there might be any association between vaccines and autism.

---

1  2  3  4  5  6  7 

Researchers in 2021 didn't find the idea so crazy after all. Reviewing evidence from independent, unrelated sources, including measurements of aluminum in the brain tissue of people with autism, they concluded that "There is a parallel rise in the association between aluminum adjuvants in vaccines for infants, and Autism Spectrum Disorder." [1]

Note: An adjuvant is an ingredient added to intentionally provoke an immune response, essentially an insult or danger the body reacts to. [2]

In 2011, IOM noted that serious side effects associated with the MMR vaccine occur mostly in children with weakened immune systems. And FDA warns that "people with weakened immune systems should talk to their healthcare provider before receiving" the MMR vaccine. [3, 4]

Caution when vaccinating people with weakened immune systems seems like smart guidance, except for one thing: ***All children have weakened immune systems at times in their lives.*** Infants receive the majority of vaccines in the first 18-months of life when their immune systems are not yet fully developed, and when the blood-brain barrier, critical to protecting the brain from toxins, is also not fully developed. Further, inflammation (again an intentional result of some vaccines) affects the blood-brain barrier. [5, 6]

Many published studies establish that reduced immune system competence is a reality for millions of American children — and for nearly every child at one time or another. [7, 8, 9, 10, 11, 12, 13, 14]

Then there are the children with an opposite condition: Their immune systems are too strong — either always or temporarily. For example, when a child is sick, the immune system becomes more active. A hyperactive immune response to a vaccine can lead to severe complications, including infection with the very disease the vaccine is intended to prevent — and shedding that disease to others. The UK

Government warns that "live attenuated vaccines should not routinely be given to people who are clinically immunosuppressed." [1]

And yet, Federal public health authorities haven't thought that's a message worth clearly broadcasting to parents, even though it likely applies to many millions of American children with autoimmune diseases, asthma and severe allergies. (One study of children with asthma found that about 10% reported "severe asthma exacerbation in the 4 weeks after immunization." [2])

Remember NIH Director Healy making this point about autism: "*Susceptible children, perhaps genetically — perhaps they have a metabolic issue, a mitochondrial disorder, immunological issue — that makes them more susceptible to vaccines plural, or to one particular vaccine, or to a component of vaccine, like mercury.*"

Since a child's immune system at the time of vaccination is clearly relevant to their risk of injury, conflicting Federal advice is truly dangerous. CDC came to the rescue in 2023 — sort of — when they provided updated guidelines for vaccinators. It's hard to expect a young, recently hired pharmacy assistant at Walgreens, CVS or RiteAid to read and apply this impenetrable guidance before giving injections to babies:

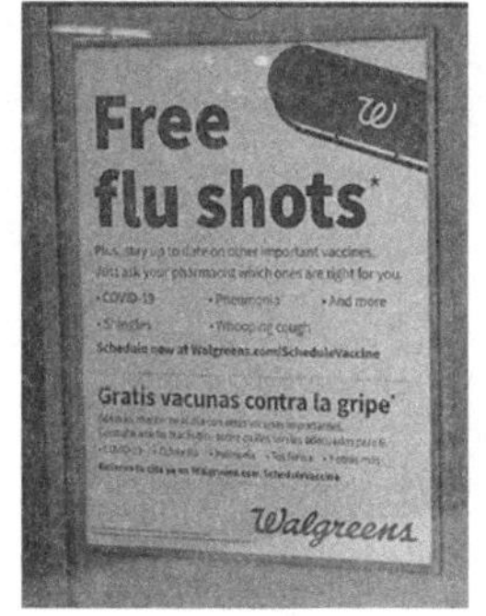

> "Primary immunodeficiencies generally are inherited and include conditions defined by an inherent absence or quantitative deficiency of cellular, humoral, or both components that provide immunity. Examples include congenital immunodeficiency diseases such as X-linked agammaglobulinemia, SCID, and chronic granulomatous disease. Secondary immunodeficiency is acquired and is defined by loss or qualitative deficiency in cellular or humoral immune components that occurs as a result of a disease process or its therapy." [3]

Not the kind of light reading a pharmacy assistant at Walmart was probably enjoying just moments before injecting a baby with an MMR vaccine, for example.

1  2  3 

"Determination of altered immunocompetence is important to the vaccine provider" because "persons who have altered immunocompetence and receive live vaccines might be at increased risk for an adverse reaction," and "administration of live vaccines might need to be deferred until immune function has improved."

Might need to be deferred?

Seems there might be a better word than *might* when warning people who might have missed the important warning that might have saved someone's life if it wasn't buried in an update that nobody might read anyway.

> "***Severe complications have followed vaccination*** with certain live attenuated viral and live attenuated bacterial vaccines among persons with altered immunocompetence. **Persons with most forms of altered immunocompetence should not receive live vaccines.**"

Good information. Negligently under-communicated by government. Embarrassingly underreported and suppressed by media. And just the kind of thing RFK Jr might ask questions about.

Speaking of information that's underreported, almost no Americans are aware of ingredients in vaccines. Historically, many things were tried, some were abandoned, some were used for a long while.

Smallpox vaccination, for example, was once accomplished using pus scraped from sores of an infected person, or scrapings from the udder of a cow. Some early vaccinations used a greasy secretion from the heels of infected horses, rubbing the smelly discharge into cuts or scratches on human skin. Those were the days of trial-and-error, sometimes combining human smallpox pus with cowpox or horsepox scrapings, thinking it would enhance efficacy. They also did vaccination by transferring pus from someone's arm to another person's arm.

Like their modern counterparts, these early experimenters didn't always get the results they hoped for, occasionally transferring terrible diseases to people. This is not the book for detailing the many gruesome and fatal outcomes, though interested readers can easily find such books, including one called *The Horrors of Vaccination*, part of a 25-book series of bad news. For balance, I'm also providing links to books by today's leading vaccine promoters, Peter Hotez, Stanley Plotkin, and Paul Offit. [1] [2] [3] [4] [5]

---

1  2  3  4  5 

Some old timey vaccine ingredients were made by steeping them for years in a mix of ox bile, glycerin, and potato slices. (Yes, really.) Over time, vaccinations evolved to include:

dried rabbit spinal cords
duck embryos
chicken blood
human bile
ground-up rat spleens
boiled pig skin

But enough about the past — we've come a long way since then. Here are some ingredients in today's modern vaccines:

Gelatin from boiled pig skin
Chicken embryo protein
Blood from the hearts of cow fetuses
Human fetus DNA fragments
Albumin from human blood plasma
Oil extracted from shark livers
Proteins from armyworm ovaries
Monkey kidney DNA fragments

There's a famous scene in Shakespeare's *Macbeth* that depicts three witches stirring a bubbling cauldron, chanting out the ingredients of their brew:

eye of newt
toe of frog
lizard's leg
tongue of dog

I don't know why that play came to mind, but let's get back to vaccine ingredients:

Formaldehyde (an established carcinogen)
Polysorbate 80 (linked to infertility)
Potassium chloride (the final ingredient administered during executions by lethal injection; obviously vaccinators use much smaller doses than executioners)
Phenol (a compound listed by EPA as a hazardous substance)
Borax/sodium borate (used in pesticides, not allowed in food, but okay for vaccines)
Monosodium glutamate (MSG)
Aluminum salts
Thimerosal/ethylmercury

Triton X-100 (also used in spermicides)

To ensure balance, I'm providing a giant dose of reassurance about how safe all those ingredients are when injected into infants and children — reassurance from the Nation's leading medical centers, well-established news organizations, prestigious universities, the Federal Government, and the World Health Organization. 1 2 3 4 5 6 7 8 9 10 11 12 13

In addition to the many websites and sources that can bolster your confidence in childhood vaccines, this chapter ends with two questions: Given all that you've read, was it really crazy of Robert F. Kennedy Jr to dig into vaccine safety? And, do you have any questions that haven't been answered to your satisfaction by those prestigious universities, medical centers, news organizations, and public health agencies?

CHAPTER EIGHT (10 MINS)

Seizures, Convulsions, Neurologic Disorders & Other Perfectly Normal Pastimes for Babies

Seizures are another adverse event some children experience soon after some vaccines, most notably MMR. For years, pediatricians have minimized these seizures, saying "It's just a febrile seizure — babies get them, nothing to worry about," even though this particular adverse effect is a major cause of emergency room visits. [1] [2]

One published paper says, "vaccine administration is the second leading cause of febrile seizures," and is a "serious concern." Why is it a serious concern? "Because it leads to public apprehension of vaccinations." [3] Another study laments that seizures which happen right after vaccination "can undermine parental confidence in vaccine safety and affect further vaccination decisions." [4]

Public health officials bend over backwards to promote the idea that seizures from vaccines are nothing to worry about. What they consider a "serious concern" is that seizures might lead parents to have reduced confidence in ***the very vaccine products that just sent their babies to the ER***. It appears that the most dreaded outcome for public health officials is vaccine hesitancy, and they know exactly what causes it: Reality.

Whenever a serious side-effect can't be denied, Federal public health officials call it rare (never quantifying the word rare). When a risk can't be denied, they say the risk is low (never quantifying the word low). When forced to admit the risks

---

1 
2 
3 
4 

are high, they say the harms are short-term. When it's an obviously serious harm, they say it's better than getting the disease one is being vaccinated against. When regression happens after vaccination, it's dismissed as something that was going to happen anyway, or had already happened — but the parents just didn't notice. And when a serious harm is absolutely, undeniably happening to children, they say it's caused by something else, or the side-effect is triggered by the vaccine, not caused by the vaccine. Or, in the case of autism, they say maybe it isn't really increasing after all.

Here's how that handwaving argument goes: Autism has always been common and all that's changed is that people recognize autism more often. Or... the definition of autism has changed and that's led to more frequent diagnosis. These myths are backwards, because while there have indeed been diagnostic changes over the years, the most recent one in 2013 tightened the criteria, making autism diagnoses ***less*** likely. And even though the new more stringent criteria was expected to reduce the frequency of autism diagnoses, autism cases still continued to increase. [1]

Another myth is that plenty of ***un***vaccinated children have the symptoms associated with autism. To explore that myth, let's look at the rates of autism diagnoses among Amish children, who typically do not receive vaccines. In 2005, when the average national rate of autism among American children was one in 125, it was one in ***15,000*** among Amish children, as confirmed by a UPI investigation. UPI also found no evidence of autism among the Amish in Indiana, Kentucky, or even Ohio, which has the Nation's largest Amish population.

The rate of autism diagnoses among American children on average is ***120 times higher*** than among Amish children. One doctor who treated thousands of Amish people in Pennsylvania reported that he'd never seen even one case of autism among his patients. Another doctor made news when he reported that he'd found an Amish patient with autism. It turned out, however, that particular Amish boy had received routine childhood vaccines. [2]

Back to seizures: Since they are so frequent after vaccination, and so often among the first signs seen by parents whose children regress into severe symptoms of autism, a lot of energy has gone into minimizing their impact. For example...

> "Vaccination appeared either to precipitate early manifestations of the condition, or to ***lead to its identification by parents***" [as in, *We didn't notice*

1  2 

> *our child had regressed mentally until that totally-unconnected seizure got our attention*]. [1]

> "There are significantly elevated risks of febrile seizures on the day of a DTP vaccine, and 8 to 14 days after the MMR vaccine..." is quickly followed by "but these risks ***do not appear to be associated with any long-term, adverse consequences***." [2]

> "The ***risk is generally low***, with estimates of febrile seizures occurring in about 1 in 1,150 to 1 in 3,000 vaccinations for MMR..." [3]

Let's do some math to see what's meant when they say a risk is "generally low." More than 90% of the 73-million children in America have received MMR vaccines, and usually two doses. If one in every 1,150 children have a vaccine-induced seizure, and two doses are given, that's about 114,000 children — and that's from just one of the vaccine types that cause seizures. They can call it rare if they want to, they can call the harms short-lived, they can call the risk low — but they can't call it misinformation. It happens, and happens to a lot of children.

One study notes that the risk of convulsion is increased in days 5–12 following vaccination, and that "the elevated risk during this time period should be communicated." It doesn't say to whom it should be communicated, but apparently not to parents making decisions about vaccines, because some might become hesitant. [4]

That study adds that seizures "need to be balanced with the potential benefit" of the vaccine. I wonder if emphasis on the benefit of the vaccine product has anything to do with the fact that all nine of the study's authors were associated with vaccine-maker Merck.

The constant minimization of risk and injury is a hallmark of CDC and FDA publications and policies. We can think of this as ***debunking in real time***.

For example, Federal public health officials haven't bothered to communicate to parents that the risk of seizures increases substantially with "concomitant multivaccination administration." [5]

English translation: The risk is higher when multiple vaccines are given at the same time, which is, alas, the norm.

> "twofold increased risk of febrile seizures after the MMRV vaccine when compared with separate MMR and varicella vaccines" [6]

---

1  2  3  4  5  6 

> increased seizure risks with combination vaccines [1]
>
> "MMRV is associated with a higher risk of seizures compared to separate vaccinations" [2]

The CDC website has been predictably misleading on the subject of convulsions: "Vaccines can cause fevers, but febrile seizures are rare after vaccination." That statement is technically accurate, because febrile means having a fever, and many vaccine-induced seizures happen *without* fever — meaning they aren't febrile seizures at all. For example, the 11-month-old baby boy rushed to the emergency room after experiencing three seizures in one morning. His emergency began after vaccination with a sharp outcry, then involuntary convulsions of his arms and legs, then his lips turned blue. But since he didn't have a fever, that didn't really count to CDC. [3]

Febrile seizures occur about four times more often in boys than in girls. (Diagnosis of autism also occurs about four times more often in boys than in girls.) [4] [5] [6]

As with many issues parents might want to consider, there is a deep rabbit-hole on the topic of vaccines and convulsions. One can find a study to say almost anything, or more accurately, one can *fund* a study to say almost anything. What you can't find, however, is a study saying that vaccine-induced seizures aren't happening. Forget about autism for a moment, and think about convulsions on their own. Since vaccine products are administered to healthy children, can seizures really be treated as a meaningless side-effect? Imagine some medical product given to healthy adults caused convulsions and loss of consciousness so often that doctors said, "Oh that's normal." Would we minimize that? Perhaps the only sense in which infants are at less risk from seizures than adults is that infants aren't standing up or driving, so they are less likely to be injured by falling or losing consciousness. Products that cause convulsions and loss of consciousness wouldn't find much of a market among adult consumers.

1  2  3  4  5  6 

"Seizure Risk after Vaccination in Children" [1]

"Vaccines and Seizures: Quantifying the Risk" [2]

"Vaccination Triggers Rather Than Causes Seizures" [3]

"Relationship of pertussis immunization to the onset of neurologic disorders" [4]

"Seizures following vaccination in children" [5]

"Nonfebrile Seizures after MMR and Varicella Combination Vaccine" [6]

"Risk of Febrile Seizures After Vaccination" [7]

"Seizures Following Childhood Vaccinations" [8]

"convulsions may be attenuated if children get MMR+V instead of the combination vaccine" [9]

"seizures following influenza vaccination" [10]

There is a very prominent physician who, like many others, chooses to withhold his name when commenting on vaccines — which tells us something about how *The Science* reacts to debate and discussion these days. "Unfortunately, because so much money has been spent to engineer the societal belief that vaccines do not cause autism," he explains, "anyone that asserts otherwise is immediately subject to widespread ridicule, to the point it's mostly a lost cause to convince medical professionals vaccines aren't always safe. In many cases, the only thing that can open their eyes is their own child being severely injured."

That exact thing famously happened to Dr. Jon Poling when his 19-month-old daughter Hannah received five vaccines all at once, containing nine different immunizations. Until then, she had been developing normally, but afterwards, not so much. Afterwards, diarrhea, loss of appetite, loss of speech, no more eye contact, and a vaccine side-effect familiar to too many parents: high-pitched screaming.

1  2  3  4  5  6  7 

8 9  10 

(Believe it or not, this side-effect is so common that there is an entire research project comparing vaccine products to see which one creates the most frequent high-pitched screaming and prolonged crying. [1])

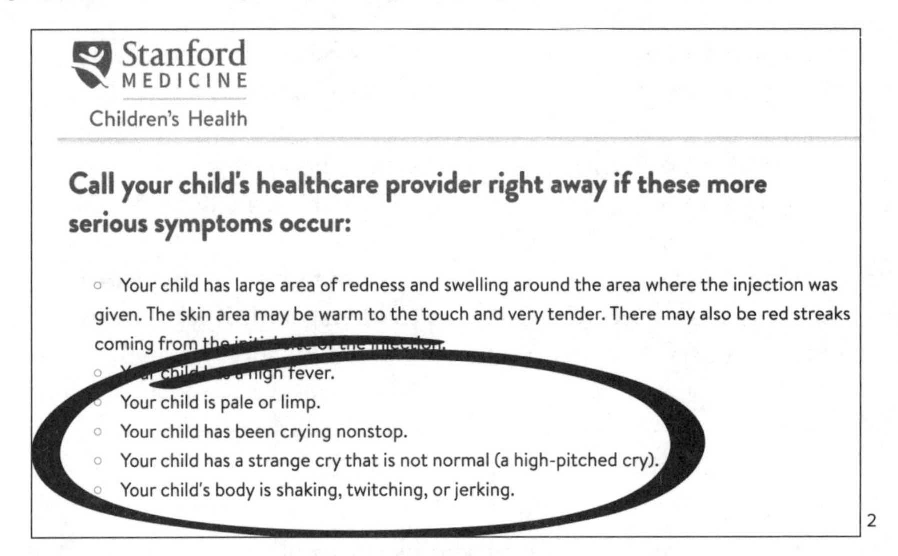
Stanford MEDICINE
Children's Health

Call your child's healthcare provider right away if these more serious symptoms occur:

- Your child has large area of redness and swelling around the area where the injection was given. The skin area may be warm to the touch and very tender. There may also be red streaks coming from the initial site of the injection.
- Your child has a high fever.
- Your child is pale or limp.
- Your child has been crying nonstop.
- Your child has a strange cry that is not normal (a high-pitched cry).
- Your child's body is shaking, twitching, or jerking.

[2]

In 2007, a Federal court agreed that vaccines played a part in Hannah Poling's autism. Accordingly, ***vaccines are the only cause of autism ever officially recognized***. I'm not saying vaccines are the only cause of autism; I'm pointing out that other than this acknowledgment, no other cause has been acknowledged by government scientists who have insisted the cause of autism is not known. CBS News did a good segment about the Poling case, however the full version is hard to find. At least for now, it can be viewed via this QR Code. [3]

After Dr. Poling spoke publicly about his daughter's case, his cancellation began, one doctor accusing him of "muddying the waters." [4] Various others weighed in to make clear that nothing was conceded other than "vaccines given to Hannah aggravated a pre-existing condition [namely, mitochondrial disease] that then manifested as autism-like symptoms." ("Autism-like symptoms" is a great phrase since autism is defined only by symptoms.) Another critic insisted this wasn't a case of vaccines causing autism, and explained without any distinction that "vaccines aggravated an underlying disease caused by bad mitochondria, and some of the symptoms Hannah showed were similar to autism."

---

1  2  3  4 

The Government parsed it best when they announced that vaccines didn't *cause* her autism, but "resulted" in it. [1]

When one critic announced that Hannah's "was a special case that could not be extrapolated to other vaccine-autism cases," Dr. Poling responded: "The only thing unique about my little girl's case is the level of medical documentation. Five to 20% of patients with ASD have mitochondrial dysfunction."

Then maybe it could happen to other children who develop mitochondrial disease, one of whom is born every 30-minutes in the US [2]

But apparently that's a crazy idea.

When it comes to the topic of vaccines, the orthodoxy leaves Americans with only two choices:

> You either automatically believe that every vaccine is safe for every child, regardless of age or how it's given…
>
> OR
>
> You are a crazy conspiracy-spouting anti-vaxxer whose skepticism is placing everyone at great risk of disease and death.

The message of this extortion is clear: ***Shut up and line up, and if some vaccine product causes harm, shut up about that, and line up for the next one***.

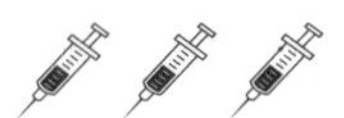

To help me make the obvious point that vaccines are not all benign and harmless, the CDC had a webpage titled "Vaccine Side Effects, Adverse Reactions, Contraindications, and Precautions." When last I visited this page, it contained 15,000 words of content we could call Fun Facts About Vaccines. But it's not fun.

For example, buried within the impenetrably dense wall of words are many important suggestions such as this about one of the polio vaccines: "vaccination of pregnant women should be avoided." There's a lot of other hard-to-find but seemingly important information CDC concealed in the (sort of) open. For children who have experienced easy bruising, frequent nosebleeds, prolonged bleeding from cuts:

---

1  2 

> "... avoiding a subsequent [MMR] dose might be prudent if the previous episode of thrombocytopenia occurred within 6 weeks after the previous vaccination."

So, if a child suffers thrombocytopenia after a vaccination, it might be prudent to avoid another vaccination with that product — you know, that product they assured parents is safe and couldn't possibly be the source of any harm.

Peppered throughout the webpage, the word death was found 35 times. Convulsion and seizure appeared 74 times; add the 12 appearances of the term *neurologic disorder* (which they explained as "e.g., a seizure"), and there were actually 86 warnings about convulsions and seizures.

"Parents of infants and children with histories of convulsions should be informed of the increased risk of postvaccination seizures." Yes, they should. But they aren't.

"In particular, they should be told in advance what to do in the unlikely event that a seizure occurs."

Wouldn't that be good?

Though I claim no medical expertise, I do speak English, and these 15,000 words were a like a dictionary that's been put through a shredder.

"These adverse events should be anticipated ***only in susceptible vaccines***..." another profoundly important phrase they never explained. It's stating the obvious to say an adverse reaction should be anticipated in children who are susceptible to the adverse reaction. But which children are those? We won't know for sure which children they are until after they have the adverse reaction. CDC brought it all home by saying that most adverse reactions to the vaccines "should be expected to occur only among the small proportion of persons who failed to respond to the first dose."

Failed to respond how? Never explained.

Another buried passage: "From 5% to 15% of vaccines may develop a temperature [fever] of greater than or equal to 103 F beginning 5–12 days after vaccination and usually lasting several days." Most people would call that being sick, but CDC called it "an excellent record of safety." Never mind that these numbers mean as many as 10-million children a year might get sick in this way from vaccine products.

When pointing out that some children will "have a neurologic event (seizure) occurring between doses," CDC didn't bother to say the seizures between vaccine doses might likely be caused by the vaccine.

I'm stopping here to protect your time and sanity, but if you find yourself with a couple of weeks to spare, you can find the old CDC website — and learn that vaccine products are not benign.[1]

All these issues above don't mean every vaccine product should be avoided; ***it's denial and ignorance that should be avoided***.

Despite all their conflicting information, despite all that CDC does not know, there is one thing CDC absolutely knew: Vaccines cannot cause or contribute to autism. That's intended to be reassuring to parents making what CDC euphemistically called "the decision to vaccinate," "the decision as to whether to begin vaccination," "the decision to administer or delay vaccination," and "the decision to give subsequent doses." Of course, none of those are really decisions at all — not if parents want their children allowed to go to school.

CDC recommends its Vaccine Schedule like an armed robber recommends you hand over your wallet. Entirely up to you.

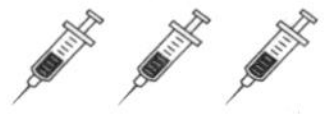

Given the situation, was it really crazy of Robert F. Kennedy Jr to dive deeper into the research? You're doing the same thing right now, and you've seen many reasons to question the credibility of Federal public health officials and vaccine makers, many reasons to be skeptical of the failed orthodoxy that has been imposed upon Americans. Among all those reasons, none is more compelling than the official insistence to keep injecting mercury and aluminum into babies.

1 [QR code]

CHAPTER NINE (15 MINS)

Mercury

"A silvery-white poisonous metallic element used in batteries and in the preparation of chemical pesticides"

... and in vaccines given to children and pregnant women.

For decades, a common ingredient in childhood vaccines has been thimerosal, which is 50% mercury. As the surge in autism (brain damage) became impossible to ignore, thimerosal was a glaringly obvious suspect. Why obvious? Because when mercury damages the human body, the main adverse effects are neurological, developmental, cognitive and behavioral, including delayed speech and loss of acquired skills. You know, a lot like severe autism. [1]

Medscape is the leading online information source for physicians. It receives much of its funding from Pharma, so Medscape is stridently pro-vaccine and pro-Pharma. Even so, Medscape advises that:

> "Mercury in any form is poisonous, with mercury toxicity most commonly affecting the neurologic, gastrointestinal and renal organ systems. Poisoning can result from mercury vapor inhalation, mercury ingestion, ***mercury injection***, and absorption of mercury through the skin." [2]

Medscape teaches that "Children exposed to mercury may experience intellectual disability," and adds, "***particularly in those moderately exposed during fetal development***." [3]

Medscape teaches that survivors of mercury poisoning, "especially those exposed in utero, often face lifelong challenges including severe developmental delays ... and persistent cognitive impairments."

---

1  2  3 

"Mercury toxicity can cause a range of neurological disorders that may persist throughout life, with particular concern for developmental impacts on children exposed ***prenatally or during early childhood***." [1]

Flu vaccines containing mercury are still given to children and pregnant women in America (more on that later). Now, ***please entirely forget about autism***, which is complicated and controversial, and focus only on neurological damage during gestation and infancy. Forget about what label is given to that damage. Focus on the fact that "even individuals with moderate exposure may experience continued cognitive impairments decades later," and the fact that exposure to mercury in any form "can lead to ***severe developmental delays and persistent cognitive impairment***."

Hold in mind that "mercury toxicity can cause a range of neurological disorders that may persist throughout life, with particular concern for developmental impacts on children exposed ***prenatally or during early childhood***."

Even the reluctant CDC acknowledges "associations between mercury exposure during development and adverse health outcomes during childhood, most notably neurodevelopment." [2]

One study found that "for each 1,000 lb of environmentally released mercury, there was a 43% increase in the rate of special education services, and a 61% increase in the rate of autism." [3]

Was concern about mercury in vaccines just a crazy idea cooked up by RFK Jr, or did he perhaps focus on the issue after studying a lot of published science, perhaps, for example, during his decade-long campaign to reduce environmental mercury in fish and waterways?

Maybe his curiosity was elevated when he learned that ethylmercury can cross the blood-brain barrier and damage the central nervous system, leading to cognitive and behavioral problems, including *autism spectrum disorder* and other developmental disorders. Maybe Kennedy became more motivated when he learned that prenatal exposure to ethylmercury can harm fetal brain development, leading to lifelong cognitive and behavioral problems.

If one is going to make the claim that exposure to ethylmercury can cause neurological injury, developmental problems, impaired language skills, and autism, well, there had better be at least one published scientific paper indicating that.

---

1  2  3 

How about 25 published scientific papers?

1. "Children are most vulnerable to mercury exposure, whether exposed in utero or as young children. Mercury affects the developing brain, causing neurological problems… delays in the development of motor skills and language acquisition, and later, lowered IQ points, problems with memory and attention deficits" [1]
2. "administration of thimerosal-containing vaccines in the United States resulted in a significant number of children developing neurodevelopmental disorders" [2]
3. "severe reactions to thimerosal demonstrate a need for vaccines with an alternative preservative" [3]
4. "A positive association found between autism and childhood vaccination uptake" [4]
5. "brain mercury levels in infant monkeys exposed to vaccines containing thimerosal" [5]
6. "Autism: A novel form of mercury poisoning" [6]
7. "The relationship between mercury and autism" [7]
8. "infants exposed to levels of mercury that exceeding EPA recommendations" [8]
9. "large amounts of thimerosal given to six patients; ***five of the six died***" [9]
10. "mercury levels showed a significant increase in infants after vaccination" [10]
11. "Mercury and autism: Accelerating evidence" [11]
12. "Blood levels of mercury are related to autism" [12]
13. "Thimerosal Induces Cell Death in Cultured Human Neurons" [13]
14. "Mercury toxicity: Genetic susceptibility" [14]
15. "Mercury concentrations in infants receiving vaccines containing thimerosal" [15]
16. "Cognitive deficit in 7-year-old children with prenatal exposure to mercury" [16]
17. "Thimerosal-containing hepatitis B vaccination and delays in development in boys" [17]
18. "Thimerosal-containing vaccines and the risk of autism" [18]
19. "Thimerosal-containing vaccines and the prevalence of autism" [19]
20. "The retention time of inorganic mercury in the brain" [20]

21. "Comparison of blood and brain mercury levels in infant monkeys exposed to vaccines containing thimerosal" [1]
22. "prenatal mercury exposure from maternal dental amalgams and autism severity" [2]
23. "Thimerosal exposure in rats mimicking exposure following childhood vaccination significantly damaged brain bioenergetic pathways" [3]
24. "Thimerosal exposure in infants and neurodevelopmental disorders: an assessment of computerized medical records in the Vaccine Safety Datalink" [4]
25. "following thimerosal exposure, the accumulation of mercury in the brain is greater than upon methylmercury exposure" [5]

> "This review found 91 studies that examine the potential relationship between mercury and ASD autism from 1999 to February 2016. Of these studies, the vast majority (74%) suggest that **mercury is a risk factor for ASD**, revealing both direct and indirect effects. The preponderance of the evidence indicates that **mercury exposure is causal and/or contributory in ASD**." [6]
>
> A systematic search of studies done over sixteen years on the relationship between mercury and autism found that among studies with public health and/or industry affiliation, 86% reported no relationship between mercury and autism. However, among studies *without* public health and/or industry affiliation, 79% *did* find a relationship between mercury and autism. [7]

Perhaps Kennedy became more concerned when he saw that many vaccines also contain aluminum (including Infanrix, Daptacel, Pediarix, Kinrix, Quadracel, Pentacel, Vaxelis, Havrix, Vaqta, Engerix-B, Recombivax HB, PedvaxHIB, and Prevnar).

---

1  2  3  4  5  6  7 

One widely published study of aluminum in vaccines posits that maybe perhaps conceivably possibly "vaccine benefits may have been overrated, and the risk of potential adverse effects underestimated."[1]

Another study asked, "Do aluminum vaccine adjuvants contribute to the rising prevalence of autism?"[2]

And another dared to place a forbidden fact right in its title, "Aluminium in Brain Tissue in Autism," and then dared to discover that the brains of children with autism contained "some of the highest values for aluminium in human brain tissue yet recorded."

Maybe Kennedy observed that when ethylmercury and aluminum were both used together in vaccines, that had a greater toxic effect than either substance alone. For example, two studies found that infant monkeys exposed to vaccines containing both thimerosal and aluminum had more neurotoxic effects and higher levels of mercury in their brains than monkeys exposed to thimerosal alone.[3][4]

But a mere two studies are hardly sufficient for believing such wild claims about aluminum in vaccines.

How about 20 studies?

1. "association between aluminum adjuvants in the vaccines and autism is suggested by multiple lines of evidence"[5]
2. "highly significant correlation between number of pediatric aluminum-containing vaccines and rate of autism"[6]
3. "Aluminum vaccine adjuvants: Are they safe?"[7]
4. "long-term persistence of vaccine-derived aluminum hydroxide"[8]
5. "exposure to aluminum from vaccines and breast milk during the first 6 months"[9]
6. "aluminum following infant exposures through vaccination and diet"[10]
7. "injection of alum-containing vaccine associated with aluminum deposits in distant organs, such as brain, still detected one year after injection"[11]
8. "vaccines increase aluminium levels in brain tissue"[12]
9. "the role of adjuvants in immune mediated diseases"[13]

10. "Human exposure to aluminium" [1]
11. "some vaccinated people present with delayed onset of cognitive dysfunction from aluminum" [2]
12. "health risk for aluminium" [3]
13. "inorganic mercury may promote neurodegenerative disorders" [4]
14. "Aluminum-induced entropy, neurological disease" [5]
15. "children may be most at risk of vaccine-induced complications" [6]
16. "some of the highest values for aluminium in human brain tissue yet recorded" [7]
17. "Levels of mercury, lead, and aluminum in autistic children are higher than controls. Exposure to toxic heavy metals at key times in development may play a causal role in autism" [8]
18. "Very high aluminium concentrations have been observed in brain samples from children with autism" [9]
19. "aluminium a powerful neurological toxicant, provokes embryonic and fetal toxic effects after gestational exposure" [10]
20. Rigorous and replicable studies have shown evidence of ethylmercury and aluminum toxicities [11]

Maybe Kennedy learned a lot when he researched and published his book on the topic, *Thimerosal: Let the Science Speak: The Evidence Supporting the Immediate Removal of Mercury—a Known Neurotoxin—from Vaccines.* [12]

Without regard to which name is assigned to a neurological disorder, it's obvious that injecting mercury into babies can have adverse effects. For this reason, vaccine fanatics had to explain away the risks. (Vaccine fanatics are those who support, favor, approve, encourage and propagandize vaccines no matter what science and reality might reveal — like CDC, FDA, and vaccine-makers, of course.) Unable to deny that mercury is toxic, they began to float a new idea: that *ethyl*mercury (the ingredient in some vaccines) is way different from other forms of mercury. Sure, it's mercury — but it's the gentle, benign, happy-baby mercury that's not worth worrying about. *Methyl*mercury, the type found in fish, that's the bad one, but *ethyl*mercury is as safe as baby powder. Oops, bad analogy because…

Though the link between baby powder and cancer was supposedly debunked (by a company paid by the talc industry), a jury disagreed and ordered Johnson & Johnson to pay $750 million to people whose cancer was caused by J&J's baby powder. Despite thousands more lawsuits lined up, Johnson & Johnson continues to insist the claims against them are "misinformation." That's a tough position to defend given that they'd been aware for decades that tests by three different labs found cancer-causing asbestos in their baby powder. [1] [2]

Much like the idea that a little bit'o mercury ain't so bad for babies, J&J sent a delegation of scientists to Washington to persuade the FDA that a little bit of asbestos ain't so bad for babies either. J&J showed the FDA that their talc "contains less than 1%, if any, asbestos." That was in 1971.

The FDA would make the IOM proud by taking decades to study what levels of asbestos should be allowed — all the while allowing it. Just to bolster your confidence in Federal public health officials, here is the headline on FDA's website today, under the banner WHAT'S NEW:

"Johnson's Baby Powder Voluntarily Recalled After Testing Positive for Asbestos"

The FDA posted this news just **48-years** after it learned about asbestos in J&J's baby powder. And how long did they wait to recall other cancer-causing baby powder brands? Just 52 years. (The recall happened in 2024.) [3] [4]

Oh well.

Rather than saying ethylmercury is as safe as baby powder, I should have said it's as safe as baby food. Oops, another bad analogy, since tests of baby food products from major manufacturers in the US found nearly all contained lead, most contained arsenic, and almost a third contained mercury. A quarter contained all three, and sometimes 10 times the limits acceptable to FDA (which are likely 10 times higher than what the limits should be, since no amount of these toxins is safe). [5]

Rather than saying ethylmercury is as safe as baby powder or baby food, I should have said it's as safe as baby formula. Oops, another bad analogy, since studies have found that almost all baby formula products contain aluminum and lead. The good news is that only 57% of samples contained arsenic. Since these products often constitute the entirety of an infant's diet, it's a shame that some contain 200

---

1 
2 
3 
4 
5 

times more aluminum than is considered safe in, say, drinking water. According to one study, most baby formulas are "not fit for human consumption," and violate the rather obvious tenet that "infant formulas should not contain anything which might endanger the health of infants and young children." [1 2]

Thankfully, IOM took on the mission of safe baby formula ingredients back in 2004, and created "algorithms to guide manufacturers in determining the level and extent of safety testing needed." So there's nothing to worry about.

By the way, who manufactures America's iconic baby formula, Similac? The pharmaceutical giant Abbott Laboratories, of course, which also makes vaccines containing ethylmercury, of course. If you can't trust Abbott Laboratories for baby formula, who can you trust?

Surely you can trust The Science, which gets us back to our main story, when vaccine fanatics were searching for some basis to support their magnificently creative idea that ethylmercury was somehow not toxic. Here's what they landed on: A study of infant monkeys indicated that ethylmercury leaves the infant's bloodstream more quickly than methylmercury — which sounds like a good result. Unfortunately, that bait and switch came with some really bad news. Though ethylmercury does indeed leave the infant's bloodstream faster than evil-twin methylmercury, it accumulates in the brain. The good news they tried to sell was decimated by the bad news:

> "... although little accumulation of mercury in the blood occurs over time with repeated vaccinations, ***accumulation of [inorganic] mercury in the brain of infants will occur***."

In fact, concentrations of inorganic mercury in the brain after thimerosal exposure can be seven times greater than that found after methylmercury-poisoning. And inorganic mercury has an infinite half-life. [3]

Further, "22 studies from 1971 to 2019 show that exposure to ethylmercury-containing compounds (intravenously, topically, subcutaneously, ***intramuscularly***, or intranasally administered) results in accumulation of mercury in the brain."

> "In total, these studies indicate that ethylmercury-containing compounds and thimerosal readily cross the blood-brain barrier, convert, for the most part, to ***highly toxic inorganic mercury-containing compounds***, which

1  2  3 

> significantly and ***persistently bind to tissues in the brain, even in the absence of concurrent detectable blood mercury levels.***" [1]

Comparing toxicity in two forms of mercury is like comparing the benefits of being shot with a .38 caliber bullet rather than a .45 caliber bullet. Most people would prefer Door #3: no bullet at all. Choosing between two mercuries is harder than trying to decide how much asbestos is okay for baby and mommy to inhale, but such is the world of debunking. For example, an Associated Press story trumpets "Big Study Finds No Strong Sign Linking Baby Powder & Cancer." Then the American Medical Association (who published the "big" study) calls the results "overall reassuring." And then the study's lead author humbly says, "This represents the best data we have on the topic." [2]

But uh oh, just weeks after the link between baby powder and cancer was debunked, a court awarded billions (more) dollars to thousands of (other) women whose cancer came from J&J's product. [3]

It's amazing that people would still claim the link between baby powder and cancer is debunked — but J&J does just that:

> "We continue to believe this was a fundamentally flawed trial, grounded in a faulty presentation of the facts. We remain confident that our talc is safe, asbestos-free and does not cause cancer."

If that J&J spokesperson were the last survivor of toxic baby powder on Earth, she'd still be repeating their press release.

Our Baby Powder Doesn't Cause Cancer is the worst slogan ever, like a restaurant chain boasting *Most of our customers aren't poisoned.*

But enough about toxic baby powder sprinkled on babies; let's get back to mercury injected into those same babies. Even if we put aside all the alarming studies (as Federal public health officials do so readily), history keeps reminding us of mercury's terrible reputation when used inside the human body.

---

1 2  3 

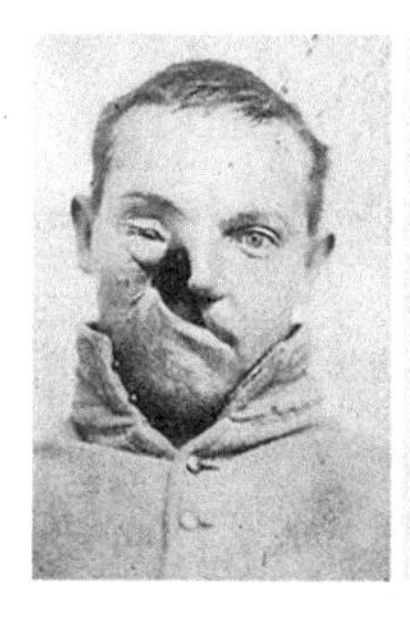   

In August 1862, 20-year-old Civil War Private Carlton Burgan was treated with a mercury compound called calomel, occasionally used back then for constipation, dysentery, fever, whatever. The mercury caused an ulcer in Private Burgan's mouth and soon enough, he lost the roof of his mouth, part of his lip, nose and right eye. A note in his medical file tells the tale: "The incompetency of the Regimental Surgeon, who prescribed the Mercury compound, caused the horrible injury." No debunking back then. [1]

Private Burgan's experience endured as a cautionary tale for decades, until it was replaced by the Minamata Mercury Disaster. That story starts in the 1950s when people in some small Japanese fishing villages noticed that their cats were behaving strangely — specifically, falling into the sea. Soon after, people, too, began to act strangely, some presenting with insanity, paralysis, and coma, followed within weeks by death.

The villagers and the cats shared a common diet —fish— and the ocean and the fish were being poisoned by the nearby Chisso Corporation's petrochemical plant. Once the problem was discovered, the company leapt into inaction, continuing to poison the water as they'd been doing since 1938. Turns out that Chisso Corporation had known all about the problem for years, ever since they'd been adding the company's mercury pollutants to cat-food as an experiment. Naturally, they kept the cats' sad experience secret — and kept dumping mercury into the ocean for another decade. It would be another three decades before the tens of thousands of people who suffered neurological damage —those who didn't die— had their day in court. If you're wondering, the Chisso Corporation is thriving. [2]

The Minamata Disaster served as a cautionary tale until being outdone by the Iraq Mercury Disaster. In 1971, seeds that had been treated with a mercury-based

---

1 2 

fungicide were imported into Iraq. Many of the sacks were marked with a skull-and-crossbones, and all featured warnings in English and Spanish. Unfortunately, since Iraqi farmers don't read English or Spanish, and since the skull and crossbones meant nothing in Iraqi culture, some of the unplanted grain was baked into bread and consumed, and some was fed to livestock. People who ate the meat were soon affected by the accumulated mercury; even small amounts caused paresthesia (numbness of skin), ataxia (lack of coordination of muscle movements), blindness, and central nervous system failure that led to death.

Researchers estimated up to 100,000 people might have suffered brain damage. (Remember, mercury in the body causes brain injury.) Thankfully, if one can be thankful for any detail in such an ugly story, most of the tainted seeds were not used. We'd know if more had been used because the seeds contained enough methylmercury to poison 3-million people. It doesn't take much. [1]

In other mercury history, old England hatmakers used the deadly metal when making felt; their obvious mental problems gave us the expression 'mad as a hatter.' You'd think mercury's long-known toxicity would discourage experts from injecting it into children, but... Trust The Science.

> Just before this book went to print, CDC's Advisory Committee on Immunization Practices finally recommended against vaccines that contain mercury. Not surprisingly, corporate news media criticized HHS Secretary Robert F. Kennedy Jr for this accomplishment. Reading this chapter, you know more facts about the effects of mercury than all the media parrots and editors put together.

1

A Conversation with Chat GPT

Gavin: Is it toxic to touch and handle mercury?

ChatGPT: Yes, **touching and handling mercury can be toxic, even in small amounts**. Mercury is a heavy metal that can be absorbed through the skin, and exposure to mercury can cause a range of health problems, including neurological damage, kidney damage, and respiratory problems. Therefore, it is important to avoid direct contact with mercury and to follow established guidelines for safe disposal of mercury-containing materials.

Gavin: Is it toxic to breathe ethylmercury, no matter how small the amount?

ChatGPT: Yes, **ethylmercury, the type of mercury found in thimerosal, is toxic when inhaled, no matter how small the amount**.

Gavin: Is it toxic to eat ethylmercury, no matter how small the amount?

ChatGPT: **Ethylmercury is toxic when ingested, no matter how small the amount**.

Gavin: Is it toxic to inject mercury into the body no matter how small the amount?

ChatGPT: Yes, **it is toxic to inject mercury into the body, no matter how small the amount**. Mercury is a highly toxic heavy metal that can cause serious harm to your health, including neurological damage, kidney damage, and respiratory problems. Injecting mercury into your body can lead to immediate and severe health consequences, including shock, kidney failure, and even death. Therefore, it is important to avoid injecting mercury or any other toxic substance into your body.

[Except in vaccines, apparently.]

For reasons poorly explained, and despite their slam-dunk debunkment of a vaccine-autism link, Federal public health authorities *e v e n t u a l l y* made a big announcement: Mercury would be removed from all childhood vaccines. You could find a thousand sources online confirming that childhood vaccines no longer contain mercury. [1]

But guess what: That wasn't true.

Here's the way CDC and FDA spun it for years: "Except for some flu shots, childhood vaccines no longer contain thimerosal," meaning, *Except for the childhood vaccines that contain thimerosal, other childhood vaccines don't contain thimerosal.*

Despite mass promotion of the idea that thimerosal was removed from all vaccines, it was actually removed from <u>some</u> vaccines — and few parents have any idea which vaccines contain mercury and which don't. Readers who are parents will confirm that no pediatrician has ever raised the issue or offered you a choice between a vaccine containing thimerosal and a vaccine free of mercury. Almost no parents know to ask a pediatrician about this topic, and sadly, very few pediatricians would know the accurate answer if asked.

Here is the accurate answer: Despite decades of being misled that mercury is no longer in vaccines, it remained in six popular vaccine products given to American children:

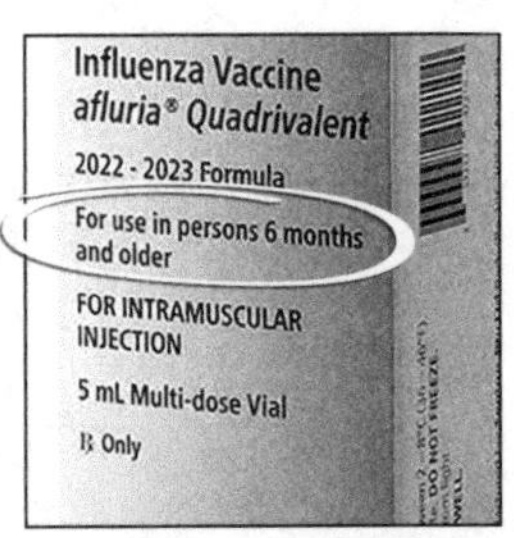

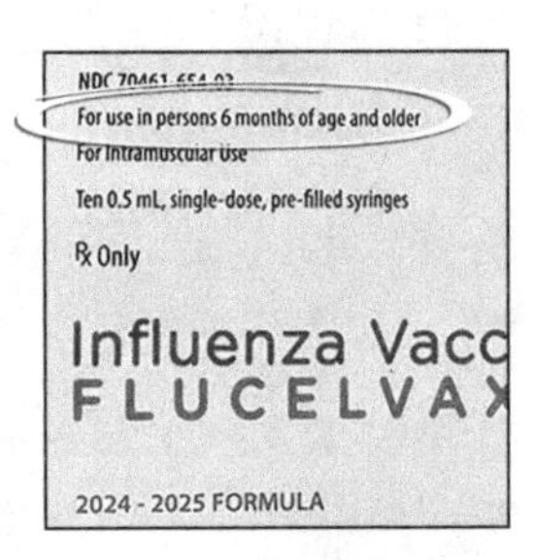

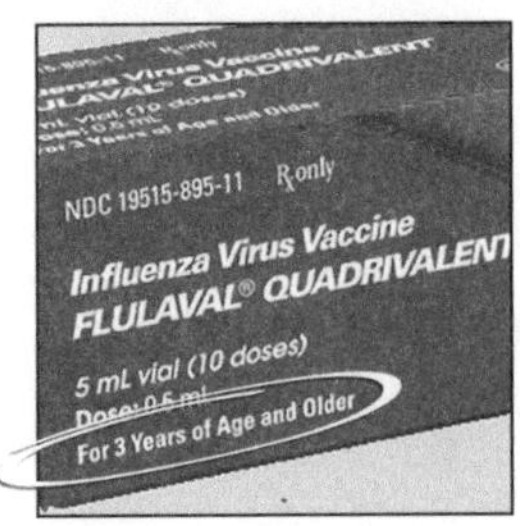

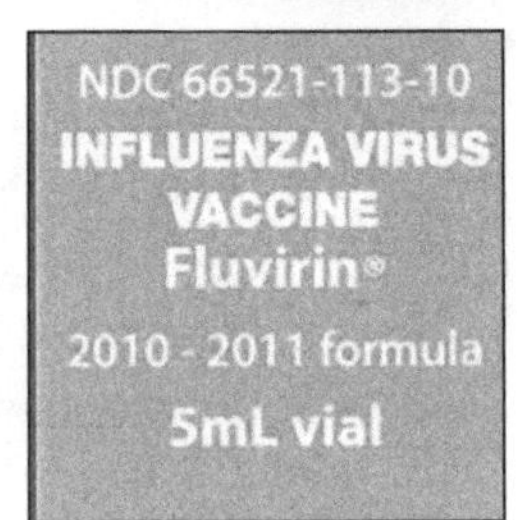

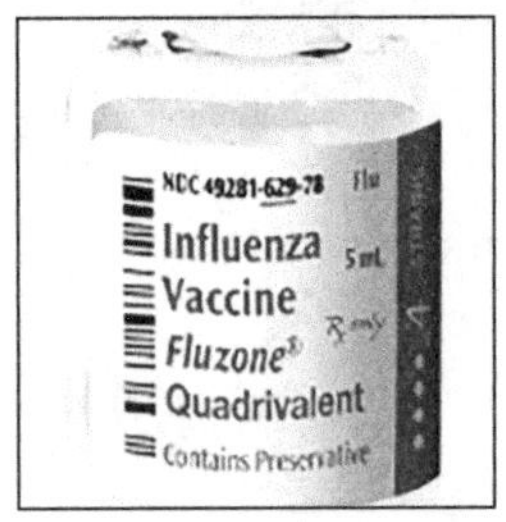

1

I paused my writing to check the CDC website, where we could find this reassuring section:

> **"Is thimerosal still used in vaccines for children?"**

Their answer is unflinchingly definitive:

> **"No. Thimerosal hasn't been used in vaccines for children since 2001."**

There is only one way to read that sentence, and it wasn't true. Lower on the same webpage, after the absolute declaration that "thimerosal hasn't been used in vaccines for children since 2001," comes this: "However, thimerosal is still used in some flu vaccines."

So, it hadn't been used since 2001, but it is still used. If a suspect answered police detectives the way CDC answered that question, they'd be in for a long interrogation. If a witness under oath testified this way, they'd be accused of perjury. The CDC, apparently, had no shame about leaving journalists and the public believing thimerosal had been removed from all vaccines given to American children.

Even though they just claimed vaccines no longer contain thimerosal, CDC continued:

> "Thimerosal is a different kind of mercury. It doesn't stay in the body, and is unlikely to make us sick."
>
> "Is thimerosal safe? Yes. Thimerosal has been used safely in vaccines for a long time (since the 1930s). Scientists have been studying the use of thimerosal in vaccines for many years. They haven't found any evidence that thimerosal causes harm." [1]

Every one of those claims was untrue, yet Federal public health officials pay no penalty for being wrong, wrong, and wrong. Speaking of lost credibility, the thousands of reports parroting the Government's lie (that mercury was removed from all childhood vaccines decades ago) led many parents to unknowingly have their children injected with mercury-containing flu vaccines. Similarly, corporate media has fails to report that mercury remains an ingredient in vaccines administered to children in India and Brazil. Oh, and Indonesia, Pakistan and Bangladesh. Plus Nigeria, Kenya, Uganda, Tanzania, and Zambia. Plus our neighbor, Haiti. Plus other countries we've treated so well, like Vietnam and Afghanistan, and oh yes,

---

1

our ally the Philippines. Literally hundreds of millions of children live in those countries, but who's counting?

Not the World Health Organization. Though they warn the world that "children are especially vulnerable" to mercury, and that when pregnant women consume fish, that "may lead to neurodevelopmental problems in the developing fetus," and that "the fetal brain is very sensitive," and that mercury can cause "mental retardation, seizures... delayed development and language disorders," the World Health Organization still says it's perfectly fine to inject mercury into babies. [1]

Quick Note: As I was completing this book in 2025, Britain's most read newspaper published this headline: "**Mercury in flu vaccine is linked to autism**." Clickbait perhaps, soon to be taken down perhaps, but the news article highlights something few Americans know: Vaccine-makers in the UK are not allowed to have thimerosal in any childhood vaccines, including flu vaccines [2] (same for Denmark, Austria, Japan, Russia, and all the Scandinavian countries).

In case you're wondering about toxicity from mercury used in dental fillings, Wikipedia has you covered: "***It has conclusively been shown to be safe.***" [3] Uh huh, except even the FDA warns that mercury fillings should be avoided by:

Pregnant women and their developing fetuses

Women who are planning to become pregnant

Nursing women and their newborns and infants

Children younger than six [4]

Oblivious to it all, vaccine-makers decided it was good business to keep on including mercury in vaccines given to tens of millions of babies. By the way, Mercury is the name of the Roman god of theft, cunning, and commerce. Coincidentally.

1

2 

3 

4 

"In science, consensus is irrelevant. There is no such thing as consensus science. If it's consensus, it isn't science. If it's science, it isn't consensus. Period."

— Michael Crichton, writer and
graduate of Harvard Medical School

"Science is a process of learning and discovery, and debate is at its heart."

— Carl Sagan

"In science, the debate is never over."

— Neil deGrasse Tyson

I'm closing this chapter with the dumbest quote I could find...

"Science is not up for debate."

— Dr. Peter Hotez

CHAPTER TEN (6 MINS)

The Swine Flu Fiasco—Which One?

In 1976, the US government told all Americans that a killer swine flu was on the way and that every man, woman, and child needed to take a new vaccine. Forty-six million Americans obediently took the shot, despite the Government knowing about a serious side effect: the terrible neurological disorder called Guillain-Barré syndrome.

Soon enough, thousands of people filed claims for serious injuries from the shot. Hundreds of others died. The mass vaccination program was stopped, and the Swine Flu Vaccines, which by the way contained thimerosal, were pulled from the market. (For comparison, 38,000 deaths and more than 2-million injuries from mRNA Covid vaccines have been reported to the Government — and those products haven't been pulled from the market. But who's counting?)

See if what follows reminds you of recent years:

The President warned the Nation about a new virus in the scariest terms possible, saying it was "similar to one that caused a pandemic in 1919, which resulted in over half a million deaths in the US"

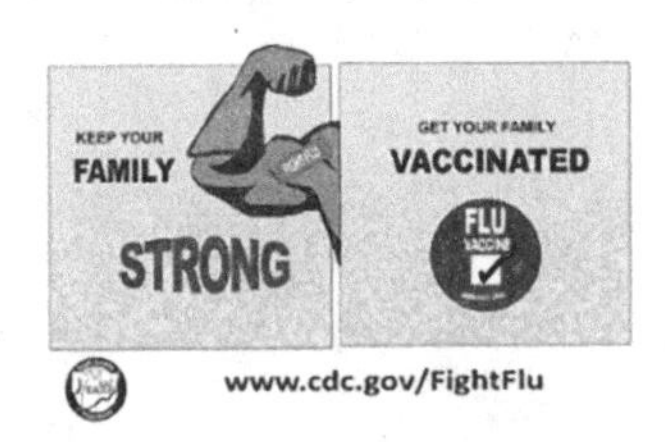

Then the President was on every news channel being injected with the new vaccine. TV commercials flooded the airwaves, claiming "The vaccines are safe, easy to take... so roll up your sleeve. Protect yourself."

How did that so-called deadly flu epidemic get started back in 1976? Some recruits at Fort Dixon, New Jersey reported symptoms of a cold. Around the same time, a soldier named David Lewis collapsed during a forced 5-mile march and died a few days later. Throat cultures sent to a lab came back as regular flu, but the CDC tested one culture and announced a different verdict: swine flu. CDC then ~~notified~~ terrified the public and promised to come up with a miracle drug.

By the way, the soldiers who'd had cold symptoms — they recovered completely, without the swine flu shot. Nonetheless, the head of the CDC devised the emergency and mass vaccination drive, and since he couldn't rely upon the single likely-unrelated death, he had to tap-dance an explanation.

> Dr. David Sencer: The rationale for our recommendation was not on the basis of the death of a single individual, but on the basis that when we do see a change in the characteristics of the influenza virus, it is a massive public health problem in this country.

Is it? Turned out it wasn't a massive public health problem after all, 'cause the pandemic never arrived. But surely CDC had very good reason to vaccinate the entire population, given the sheer number of swine flu cases around the world at that point. What was the sheer number?

> Dr. Senser: There had been several reported, ***but none confirmed***.

No 1976 swine flu outbreaks were ever found anywhere. For clarity, the entire mass vaccination plan (to inject every man, woman and child in America) was based upon zero confirmed cases of swine flu. But zero is more than enough for CDC to build a full-scale pandemic panic — and then assure everyone that Vaccine X53A was coming to the rescue.

The CBS News show *60 Minutes* asked if X53A was ever field-tested.

> Dr. Sencer: I can't say. I would have to—
>
> 60 Minutes: It wasn't.
>
> Dr. Sencer: I don't know.
>
> 60 Minutes: I would think that you're in charge of the program.
>
> Dr. Sencer: I would have to check the records. I haven't looked at this in some time.
>
> 60 Minutes: Did anyone ever come to you and say, *You know something, fellas; there's the possibility of neurological damage if you get into a mass immunization program*?

> Dr. Sencer: No.
>
> 60 Minutes: No one ever did?
>
> Dr. Sencer: No.

But that warning had, in fact, been sounded right inside CDC by Dr. Michael Hattwick, the man whose job it was to assess whether X53A might be harmful.

> 60 Minutes: Did you know ahead of time, Dr. Hattwick, that there had been case reports of neurological disorders, neurological illness, apparently associated with the injection of influenza vaccine?
>
> Dr. Michael Hattwick: Absolutely.
>
> 60 Minutes: You told your superiors, the men in charge of the swine flu immunization program, about the possibility of neurological disorders?
>
> Dr. Hattwick: Absolutely.

When Dr. Sencer denied this to *60 Minutes*, correspondent Mike Wallace showed the CDC Director a CDC report that listed neurological complications as an adverse effect of the Swine Flu Vaccine. Dr. Sencer fell back on a popular standby in public health: an invented scientific consensus.

> Dr. Sencer: I think the consensus of the scientific community was that the evidence relating neurologic disorders to influenza immunization, that they did not feel that this association was a real one.

This means the scientists who work for us will say so. Whatever the risks, CDC developed a giant advertising plan to urge Americans to take the shot. One ad reported that "The swine flu vaccine has been taken by many important persons," listing as examples Henry Kissinger, Elton John, Muhammad Ali, Edward Kennedy, and TV star Mary Tyler Moore.

But Mary Tyler Moore, for just one example, had not taken the Swine Flu Vaccine, even though her doctor had initially thought it might be a good idea. "I was leery of having the symptoms that sometimes go with that kind of inoculation," she prophetically said.

> 60 Minutes: Have you spoken to your doctor since?
>
> Mary Tyler Moore: Yes, and he's delighted that I didn't take that shot.

Some ads used famous people to push the vaccines, but as in recent years, CDC decided fear was the best medicine to use in its commercials. One showed people happily greeting each other, kissing and hugging, oblivious to the invisible menace lying in wait at the end of the commercial.

> "Joe brought it home from the office. He gave it to Betty and one of his kids, and to Betty's mother. But Betty's mother went back to California the next day. On her way to the airport, she gave it to a cab driver, a ticket agent, and one of the charming stewardesses. At school, Joe's kid gave it to some other kids. And Mrs. Merrill got it and gave it to her husband. In California, Betty's mother gave it to her best friend, Dottie. But Dottie had a heart condition, and she died. But before she died, Dottie gave it to her girlfriend, the mailman, the paperboy, and the vet when she went to pick up her chihuahua. If a swine flu epidemic comes, this is how it could spread."[1]

After the 1976 vaccine product was recalled, CDC stopped talking about the swine flu... but not forever. The whole thing happened again in 2009. Yes, another mass vaccination campaign for swine flu, this one led by Dr. Anthony Fauci. He aimed to get every American injected, including babies right down to 6-months old. In the end, he had to settle for only 70-million Americans and only 26-million children getting those 2009 swine flu vaccine products (most of which contained thimerosal).[2]

As usual, public health officials in 2009 failed to adequately inform Americans about the risks of the vaccines. Care to guess what the risks were? Guillain-Barré Syndrome again, and more. Facial palsy, convulsions, and plenty of deaths. But not enough deaths to influence Dr. Fauci's video message to Americans at the time:

> "The track record for serious adverse events is very good. It's very, very, very rare you ever see anything associated with the vaccine that's a serious event."[3]

1

2 

3 

So many *verys* must mean it was true. Only it wasn't. More than a thousand Americans likely died from that vaccine, meaning they missed out on many years of future reassurances from Dr. Fauci.*

The UK government reported that the 2009 Swine Flu Vaccine caused narcolepsy, and also deaths. Like most of the Swine Flu Vaccine products, the main one used in the UK contained thimerosal. UK officials reassured the public that "The vaccine has been thoroughly tested." Except, that wasn't true. [1]

But vaccine safety was not the reason Dr. Fauci's 2009 mass vaccination drive petered out before he got every last American. The public lost interest because people just weren't scared enough. That lesson was well learned and remembered in 2020, when the next mass vaccination campaign was launched. By then, Dr. Fauci and CDC and FDA and corporate media and Big Pharma had together perfected how to scare most Americans into compliance.

* The Government's Vaccine Adverse Event Reporting System (VAERS) contained reports of only 40 deaths associated with the 2009 Swine Flu Vaccine. However, VAERS is known to capture fewer than 1% of the actual adverse events, meaning thousands of deaths associated with the 2009 Swine Flu Vaccines would have gone unreported to the Federal Government. Many studies, including at least one commissioned by the Government, have demonstrated substantial underreporting in the VAERS system. [2] [3] [4] [5] [6] [7]

1  2  3  4  5  6  7 

CHAPTER ELEVEN (14 MINUTES)

Vaccines Are the Greatest Idea Ever Conceived, for Real

Most people would define a vaccine as *a benign and harmless injection that reliably prevents infection and transmission of a disease that children are likely to be exposed to, and which, in the absence of vaccination, could be fatal or debilitating.*

That's the greatest idea ever. A safe injection that doesn't harm anyone and prevents a child from getting and spreading a terrible disease. No sane person could oppose that idea. But does that definition actually apply to every vaccine product that's been marketed in the US?

Let's start with benign. We have seen with clear eyes that no vaccine is 100% harmless to all recipients. That's obvious and has been acknowledged by Justices Sotomayor and Ginsburg in their Supreme Court opinion on the topic:

> *There are some products which, in the present state of human knowledge, are quite incapable of being made safe for their intended and ordinary use. These are especially common in the field of drugs. An outstanding example is the vaccine for the Pasteur treatment of rabies, which not uncommonly leads to very serious and damaging consequences when it is injected.*

The Justices were commenting on the National Childhood Vaccine Injury Act of 1986, a law that eliminated all financial liability for companies that make what are acknowledged to be "unavoidably unsafe vaccines."

There are many other products that are unavoidably unsafe, including cars and airplanes, for example. However, makers of cars and airplanes didn't lobby for a new law to absolve them of all product liability. That's because makers of cars and planes weren't (and aren't) overwhelmed by successful lawsuits. Conversely, since the 1970s the vaccine industry faced a swelling wave of lawsuits driven by severe adverse reactions, often neurological damage. Mounting legal costs and large jury awards inspired the industry, notably Wyeth (now Pfizer), to actively lobby the

Reagan administration and Congress for liability relief. They made the case that their products sometimes harm people, and that they can't help it. [1 2]

Anyway, some vaccines cause serious injury to some people, including lifelong disabilities and death. *Not* benign.

Next, no vaccine provides 100% immunity from the disease it is designed to prevent. Some (e.g., Covid, DTaP, Tdap vaccines) don't prevent infection or transmission at all.

It is also worth considering that nearly all diseases for which healthy American children are given vaccine products have nearly 100% survival rates for those children, meaning these diseases are never the instant-death bogeymen they are promoted to be. Vaccines are many different products that pose different risks, marketed to address diseases that also pose different risks. Here are some estimates people can argue about:

Hepatitis B:

survival rate for a healthy American infant is effectively 100%

99% of healthy American infants who contract Hepatitis B will not experience any symptoms at all. [3]

Rotavirus:

survival rate for a healthy American child is effectively 100%

Nearly every American child contracts rotavirus at some point, whether or not vaccinated. Many millions of American children (30–40%) do ***not*** get the rotavirus vaccine recommended by CDC (three injections in the first 24-weeks of life), and yet survive. [4]

Haemophilus influenzae type b (Hib):

survival rate for a healthy American child is effectively 100%

There appear to be zero Hib deaths among American children, a fact credited to vaccines. However, vaccines cannot explain why many millions of American children who are ***not*** vaccinated against Hib also do not die from Hib. CDC recommends 3–4 Hib vaccines administered to babies in the first 12-months of life. [5 6]

Mumps:

survival rate for a healthy American child is effectively 100%

Mumps deaths do not occur in the US, not even among the many millions of American children who are not vaccinated for mumps. CDC acknowledges that "the majority of cases and outbreaks occur among people ***who are fully vaccinated***." It's a little more than the majority — more than 90% of the mumps cases in America are in people vaccinated against mumps. That's not much of an endorsement, and neither is this: "Experts aren't sure why vaccinated people still get mumps." Maybe it's just that the vaccine doesn't work so well. Good thing mumps typically resolves on its own in about a week. [1 2]

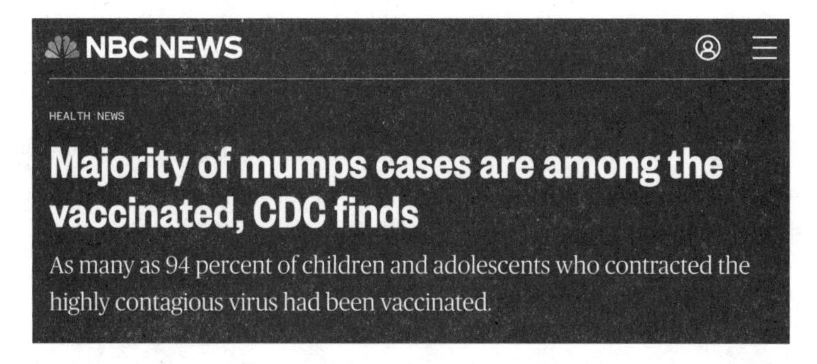
NBC NEWS

HEALTH NEWS

Majority of mumps cases are among the vaccinated, CDC finds

As many as 94 percent of children and adolescents who contracted the highly contagious virus had been vaccinated.

Polio:

Last year, no children anywhere on Earth died from polio. Perhaps surprisingly, the vast majority of people who get infected with poliovirus will not experience any symptoms at all. A small number will have flu-like symptoms that last 2 to 5 days, then go away on their own. Yes, I am describing polio, and I'm using the CDC's exact words to do it. The infamous consequence associated with polio is paralysis, which occurs very rarely, about half of one-percent among those with symptoms. And about 50% of those people recover fully. I know it's all hard to believe, and that's why you have the source citations, including this one ("In spinal paralytic poliomyelitis, approximately half of cases recover fully"). [3]

For a long while, there hadn't been a case of polio in the US since 1983. Some readers are likely yelling, "That's because of the polio vaccine!" I hear you. ***But that cannot account for the nearly 2-billion people on Earth (including millions in the***

1  2  3 

US) who are not vaccinated against polio. They also aren't dying of polio. And unfortunately, the most recent instance of polio in the US was found to be ***vaccine-derived poliovirus***. That's also because of the polio vaccine.

In fact, almost all polio discovered today is ***vaccine-derived polio***. Last year, there were 536 cases of polio paralysis on Earth, and 97% of those were ***vaccine-derived polio***. According to CDC, almost all those cases were in the Congo, Nigeria, Yemen and Somalia. [1] (***Vaccine-drived poliovirus*** has also been found in wastewater in Spain, Poland, Finland, Germany, and the UK.) [2]

Though the facts above are facts, and thus ought not be controversial, they are forbidden facts. Otherwise, you'd have heard that almost all polio cases have zero symptoms, and almost all cases with symptoms resolve on their own in a few days. Public health officials often express that same truth in a more alarming way, describing polio as "incurable," like measles is "incurable." What they mean is that there is no Pharma product or treatment that cures the disease, however if you wait a week, time itself will cure nearly 100% of those few people who have symptoms. I am not minimizing the possible though very remote risks of polio; rather I am rightsizing the possible *though very remote* risks of polio.

Even as I write this, CDC has been focusing extra attention on a virus called HMPV: "There is no vaccine or cure for the virus at this time." Less emphasis is placed on the fact that HMPV symptoms are like the common cold, improve with rest, fluids and over-the-counter medication — and it's typically over in a week. But take no comfort, for "there is no vaccine or cure for the virus." [3]

In merely sharing these facts, I am not minimizing any disease. On the other hand, public health officials and vaccine makers vigorously ***maximize*** every disease, promoting the horror of any disease for which they have a product.

A CDC-to-English dictionary might be helpful when you read alarming statements such as this: "The onset of paralysis usually occurs 7 to 21 days after infection." The word *usually* is very misleading, since the onset of paralysis *usually* ***doesn't occur at all***. A more accurate way to share that information would be to say that in the *very unusual event that paralysis occurs from polio (which is less than 1% of cases), the onset of that incredibly rare paralysis usually occurs 7 to 21 days after infection.*

---

1  2  3 

I said above that public health officials and vaccine makers vigorously promote the horror of every disease for which they have a product — and oh boy, have they got products.

In the case of polio, for example, there have been many different vaccine products, and there are currently six on the market in the US.

IPOL	Pediarix
Pentacel	Kinrix
VAXELIS	Quadracel

These six products are not identical, obviously, or else there wouldn't be six of them. They have different ingredients, different dosages (some require four doses, some three), and they have different safety profiles. Vaxelis causes high fever more often than its competitive products. Vaxelis, Kinrix, and Pediarix cause decreased appetite and drowsiness, and only Vaxelis and IPOL specifically list vomiting as a side effect. [1]

IPOL appears to have the fewest adverse effects, and yet it's the one used least often. Go figure. My overall point is that there is a decision to be made about which of the six will be injected three or four or five times into one's baby — and the same is true for all childhood vaccines. Somebody needs to make a decision. The vaccine makers, the Government, and the medical-industrial complex will always recommend the maximum number of vaccine products and doses for every child — and will promote the most frightening descriptions of each malady. Here for balance is more balance:

Pertussis:

survival rate for a healthy American child is effectively 100% [2] [3]

Tuberculosis:

the deadliest of all infectious respiratory diseases, killing hundreds of thousands of children every year. And yet, the TB vaccine (called BCG) is not given in America, even though it's been around for more than a hundred years and is used in almost every other country on Earth. Some speculate the BCG vaccine isn't given in America because tuberculosis is relatively rare, with only 9,366 US cases in 2023. But that same point is far more

1 
2 
3 

applicable to tetanus (fewer than 30 cases in 2023) or polio (only one case that year) or diphtheria (zero cases that year). [1]

I included tuberculosis on this list to point out that ***where someone lives*** can be important when comparing the impact of diseases to the risk/benefit of vaccine products. There's a more perplexing aspect of this public health decision in the US The BCG vaccine has far and away the most beneficial health results of any vaccine, with emerging evidence indicating it can:

- afford protection against unrelated respiratory infections and sepsis, particularly in children
- lower the incidence of bacterial and viral infections
- reduce the risk of allergic conditions like asthma, eczema, and allergic rhinitis
- improve autoimmune conditions like type 1 diabetes and multiple sclerosis
- help treat bladder cancer by destroying cancer cells and preventing tumor recurrence [2]

And <u>*this*</u> is the vaccine that isn't routinely given in the US Go figure.

Tetanus (lockjaw):

survival rate for a healthy American child is effectively 100%

Also, tetanus is not a communicable disease, and is very, very rare. In all of the United States, ***over a 10-year period***, there were 13 tetanus deaths, all elderly people. In a whole decade, only 51 young people in America contracted the infection, with no deaths. Important Note: A tetanus toxoid vaccine can be given at the time of a deep puncture wound, which is when the very rare bacterium might enter the body. Even if a person has had a tetanus vaccine within recent years, doctors will almost always recommend getting another vaccine as part of treatment for a deep puncture wound. [3]

Measles:

survival rate for a healthy American child is effectively 100%

Though as many as 9-million American children are not vaccinated against measles, deaths from measles are extremely rare, and most years go by with zero measles deaths. Since the measles vaccine is not fully protective, the

---

1  2  3 

few cases of the illness (typically fewer than 100 per year) occur among both vaccinated and unvaccinated children. [1]

Terrible-sounding statistics such as "5% of American children hospitalized for measles had infections in their lungs," are less alarming when we realize that 5% amounts to somewhere around three children in all of the US in all of 2024.* [2] [3]

Chickenpox:

survival rate for a healthy American child is effectively 100%

"Chickenpox is usually self-limiting and will resolve by itself within 7–10 days." [4]

Hepatitis A (HepA):

survival rate for a healthy American child is effectively 100% [5]

Covid-19:

survival rate for a healthy American child is effectively 100% [6]

Pneumococcal disease:

survival rate for a healthy American child is close to 100%

The most serious result of pneumococcal disease is known as IPD, and "most fatal IPD cases are currently not vaccine-preventable." [7]

And yes, I still hear the yelling. And yes, I know (and here restate) that adverse vaccine reactions are rare. And yes, being ill is worth avoiding — but not at all costs. If you want to be reminded about the terribleness of these particular diseases (while simultaneously hearing nothing about childhood diseases for which there is no vaccine), there are thousands of websites to remind you. I'm seeking to allow reality to balance the medical establishment's suppression and exaggeration of reality. Though the diseases are always vilified and the vaccines always glorified, even IOM acknowledges that.

"immunization is not without risks… It is well established, for example, that the oral polio vaccine can on rare occasion ***cause*** paralytic polio, that

* [From CDC: Total measles cases in the US in children under 5 in 2024 = 120. Hospitalized = 62 children. 5% of 62 = 3 children.

1  2  3  4   5   6  7 

> some influenza vaccines have been associated with a risk of Guillain-Barré syndrome, and that vaccines sometimes produce anaphylactic shock."[1]

Vaccines sometimes produce death, too, but apparently, IOM didn't find that worth mentioning. Having shared the above survival rate percentages that apply to children who experience these diseases, it is also important to note that most American children are not likely to be exposed to any of these diseases — and certainly not 100% likely — and certainly not likely to be exposed to all these diseases. Since no vaccine is 100% safe and no vaccine is 100% effective, and no disease is 100% catastrophic, and no child is 100% certain to be exposed or infected, and some vaccines harm some children, it makes sense to have very low tolerance for risk from the products themselves. It boils down to parents deciding what percentages work for them.

For example, imagine that Covid is not a serious infection for healthy young people. Imagine that Covid mRNA vaccines do not prevent Covid infections or transmission in any event. And imagine that sometimes, though rare, the Covid vaccine causes some children to die from sudden cardiac failure during their sleep or during play, and less rare, causes serious cardiac injury to young people. (All true, by the way, and no imagination required.)[2] [3] [4] [5] [6] [7] [8] [9] [10] [11] [12]

Staying a moment longer with just the mRNA vaccines, Pfizer's own safety data on their Covid product (which they brand as Comirnaty) reveals:

> "The safety and effectiveness of COMIRNATY in individuals younger than 12 years of age ***have not been established***."
>
> "Available data on COMIRNATY administered to pregnant women ***are insufficient to inform vaccine-associated risks in pregnancy***."
>
> "Postmarketing data with authorized or approved mRNA Covid-19 vaccines demonstrate ***increased risks of myocarditis and pericarditis***, particularly within the first week following vaccination. For COMIRNATY, the observed risk is highest in males 12 through 17 years of age... ***some cases required intensive care support***. Information is not yet available about

> potential long-term sequelae." (Sequelae is a word used because few people know that it means pathological condition, injury, trauma.) [1]

Unlike cardiac injury and blood clots, the mild adverse effects that Pfizer acknowledges include things like severe headache, severe fatigue, chills, high fever, nausea, vomiting. Call me crazy, but I used to associate these things with being sick. Since this product is given to healthy children, it seems wise to be sure the intervention is worth it.

Dr. Paul Offit concluded that getting the mRNA Covid booster would not be worth the risk for the average healthy 17-year-old boy, and advised his own son against getting a third dose. [2]

Parents can ask themselves what risk percentages work for them when deciding whether or not to give their child or infant a Covid mRNA vaccine (or 2 or 3 or 4 or 5 or 6 or 7 Covid vaccines like CDC has recommended). In other words, only parents can decide how much risk of very unlikely outcomes is worth how much vanishingly improbable benefit.

For public health officials, it's a math exercise with many zeroes after each decimal point, but for individual families, it cannot be mere numbers. For example, a healthy 12-year-old boy who has recently had his third Covid mRNA vaccine might gain the benefit of a slightly milder case of Covid. At the other end of the spectrum, he might die from vaccine-induced inflammation of the heart. (Oh, but that's so rare!) If there's one good thing about mRNA vaccines for children, it's that the decision is easy.

While CDC previously recommended that infants 6-months and older receive ***three*** mRNA injections (to start), the Government made no case whatsoever for why such a policy makes sense for babies. Accordingly, parents have to do the calculus for their children, and in 2024, ***97% of American parents decided to not follow CDC's recommendation.*** [3]

Are they all bad parents who don't care about their children and don't care about other people? Not likely, any more than parents who gave their infants three mRNA vaccines are bad parents. All are doing what they feel is right for their children, and all are influenced to some degree by fear of the diseases, fear of taking chances, fear of being wrong, fear of social judgement.

---

1 
2 
3 

Are parents who decide against vaccines too dumb to understand *The Science*? Do they think they know better than the experts? No and No. Just like parents who choose to give every recommended vaccine, these parents are doing what they feel is best for their children. Whatever their individual reasons for hesitancy about some vaccines, they are not alone...

> 23,000 people in 23 countries surveyed; decreased intent to get Covid-19 booster;
> Lazarus et al. (2023)
>
> Cleveland Clinic study found trust in healthcare providers is key determinant of vaccine hesitancy;
> Oluwatosin et al. (2024)
>
> 30% of Italian students are vaccine hesitant;
> Baccolini et al. (2021)
>
> Healthcare workers 50% undecided, 28% refused vaccination;
> Fares et al. (2021)
>
> High-income countries have vaccine hesitancy rates as high as 77%;
> Sallam (2021)

Stanford Medical Center informs parents:

> "As with any medicine, vaccines may cause reactions... Serious reactions are rare. But they can happen. Your child's healthcare provider or nurse may discuss these with you before giving the shots. The risks for getting the diseases the shots protect against are higher than the risks for having a reaction to the vaccine."

It's correct that serious individual vaccine reactions are each rare, but not as rare as serious complications from the diseases. Why? Because no children are exposed to all the diseases, few are infected with any of the diseases, and even fewer have any symptoms — but ***millions of children get all the recommended vaccines***.

Stanford says your child's healthcare provider ***may*** discuss the possible serious reactions, which also means they may ***not***, which turns out to be the case most of the time. The "risks for getting the diseases" sounds like they mean the risks ***from*** the diseases, as in the dangers — but that's not what they are saying. Before most parents could think to actually compare the risks to the risks, Stanford distracts with an appeal to our hearts:

> "Children may need extra love and care after getting immunized. The shots that keep them from getting serious diseases can also cause discomfort for a while."

Stanford's extra love prescription won't be enough for the kids who experience those serious reactions the healthcare provider ***may*** have warned about — but didn't. As we assess information from Stanford or any other medical center, remember these are businesses that make money administering vaccine products. Stanford has a thriving pediatric medical group that "believes all children should receive the recommended vaccines," unsurprisingly. On their homepage is a smiling just-vaccinated child proudly displaying a bandaid. Next to the child is the link to "Schedule Your Appointment Today." Like all medical center businesses, this one promotes and sells vaccine products. [1]

> **caveat emptor** *(noun)* — The axiom that the buyer alone is responsible for assessing the quality of a purchase before buying. A commercial principle that without a warranty the buyer takes upon himself the risk.

I gladly restate that adverse reactions to most vaccines are rare, but the promotional way medical institutions communicate about vaccines makes it difficult to compare the possible benefit of a particular vaccine product to the possible adverse reactions. When accurately informed, human beings are quite capable of making such comparisons by applying what we can call the Law of Unlikely & Infrequent Events (LOUIE).

You already live your life assessing the LOUIE formula, deciding which events are so unlikely that they are effectively irrelevant. For example, a burglar could arrive at your house by helicopter and core through your roof, but you've applied LOUIE and decided to do nothing to reduce that particular risk. On the other hand, because a burglar coming in through the front door is reasonably likely, you keep that door locked.

In our minds there is a list of things we want to avoid or prevent. We decide which precautions are reasonable and which precautions are so expensive (in time, energy, money, risk) that we don't pursue them. If there's an emergency contact

---

1 

list in your home, the names and numbers reflect your family's assessment of likely hazards. Your list probably doesn't contain the phone number of the US Nuclear Emergency Search Team, the number you'd call if you thought a nuclear bomb was hidden in your neighborhood.

Your list has limits — because it has to. Every year, there are new offerings on the growing menu of vaccine products. Some might be worth taking, some might not; some decisions might be obvious, some might be nuanced. Do you prefer the decisions be left to vaccine-makers, government officials, news media — or to parents?

Whoever decides, each vaccine product calls for its own risk/benefit assessment, and its own comparison to the risks of the disease. That's always a matter of odds and probabilities, combined with people's different approaches to risk. Earlier in the book, I suggested an exercise to clarify what risks and benefits would work for you when the decision is yours. Fill in the blanks that reflect your views and feelings at this point:

- I would give my child this particular vaccine if it prevents infection and transmission of the disease
  ___ not at all
  ___ very rarely
  ___ most of the time
  ___ all of the time
- and if the disease would otherwise cause death or life-altering health problems
  ___ for zero healthy American children who get the disease
  ___ for almost no healthy American children who get the disease
  ___ for most healthy American children who get the disease
  ___ for every healthy American child who gets the disease
- and if my child is likely to be exposed to this specific disease
  ___ effectively never
  ___ very rarely
  ___ occasionally
  ___ often
- **and only if the vaccine itself has caused a child's death or life-altering injury**
  ___ no more than 5 times.
  ___ no more than 25 times.
  ___ no more than 100 times.

__ no more than 1,000 times.
__ no more than 10,000 times.

Based upon these selections and other considerations that reflect their personal feelings about risk and benefit, parents can then vaccinate their children against the diseases that concern them most, using the vaccine products that inspire confidence the most. That might be every vaccine, or some.

We all tend to believe what we believe because it feels better and is more convenient than the uncomfortable alternatives. Many of our beliefs are formed socially rather than scientifically. This book asks that we step outside our tribes (conservative/liberal), and outside the tropes (safe & effective), and outside our masters (mandates) — and evaluate confirmable information. ***Information is neutral.***

It's clear that no parent would let a stranger inject something unknown into their child — and yet millions of parents do exactly that at the pharmacy. To do so, we must blindly trust the pharmacy chain to effectively screen and hire and train and supervise the employees who prepare and administer injections to our children. Parents are also asked to blindly trust Federal regulators and public health officials, medical businesses, doctors and pharma companies.

Since few people know much about the ingredients in the injections, millions of parents turn the whole matter over to authority figures and institutions in Washington DC (strangers). These powerful officials are impatient when people ask questions. They silence those who challenge their opinions (including doctors and scientists), and they invent strategies to avoid scrutiny and prevent choice. Their most effective strategy by far is Fear, because when people are afraid, they'll take any train that's leaving the station — even if they don't know where it's going.

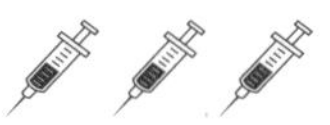

My purpose in these pages is not to persuade anyone that some vaccines can contribute to autism in some children, or that the combination of many vaccines given together might injure some children, or that other factors might contribute to autism in some children. Rather, my purpose is to show that the topic is not so black and white as people have been led (and nearly forced) to believe. The word vaccine is like a brand that most people just accept to mean safe, effective, miraculous, and benign. But vaccines are many different products, with different ingredients, different levels of testing, different benefits, and different risks — given to children with different physiologies and vulnerabilities. Vaccines are not one thing.

Like other products, each one is developed, licensed, branded, marketed, and sold for profit. But unlike other products, we are expected to simply accept all manufacturers' claims and ask no questions. With more and more vaccine products being pushed on (and into) our children, it's going to require more thinking and choosing and deciding.

Some people might accept hand-waving arguments about "scientific consensus," or believe, for example, that hundreds of studies have debunked the vaccine-autism link. But that... is... not... true — and the people promoting it are not telling the truth. If you prefer information without bias, it's not likely to come from paid debunkers, pharma companies, medical businesses, or captured government agencies — unless you've always found those sources to be credible, honest, objective, and correct. For readers who'd like to dig deeper into the pros and cons of each vaccine, there's a well-researched article entitled "What Are the Pros and Cons of Each Vaccine." [1] There is also a superb and enjoyable book about the history of vaccines called *VACCINES: Mythology, Ideology, Reality* by John Leake & Peter A. McCullough M.D, MPH. [2]

One solution for all these challenges is for Federal public health authorities and Pharma to regain the public's trust. While we're waiting for that, there will be new vaccine products for new diseases and conditions, and American parents will have to decide which ones make sense for their children. It's not likely to be no vaccines — and it's not likely to be every vaccine.

---

1  2

CHAPTER TWELVE (9 MINS)

Childhood Vaccines Have Saved 150-Million Lives

Everybody knows that vaccines have saved millions of lives. Like gravity, it's obvious, and like gravity, there's little reason to ask how we know it. Nonetheless, since there are always oddball skeptics asking buzzkill questions, let's follow The Science and put this to rest.

There are many scientific papers, studies, and reports that confirm vaccines have saved 150-million lives (154-million to be exact). We'll start with the study done by the Imperial College London and published in the prestigious journal *Lancet*:

> "Vaccination has averted 154 million deaths, including 146 million among children younger than 5 years of whom 101 million were infants younger than 1 year."

Next there's the PATH report (*Vaccines: 50-years of Progress*) that explicitly states "vaccinations have saved 154-million lives since 1974."

Maybe we shouldn't count the PATH report because it's based entirely on the first study above.

That still leaves the Global Health Security Agenda Consortium, which also confirmed 154-million lives saved. Oh, maybe we shouldn't count this one either because it's also based entirely on the first study above.

Nonetheless, we can rely upon the report from the School of Hygiene and Tropical Medicine, which concluded that 154-million lives (101-million of them infants) were saved by childhood vaccines. I guess we can't count this one either because it too is based on the first study above. Sorry.

We can instead draw upon the major article in the respected journal *Nature*, which states that "global vaccination efforts have saved 154-million lives in 50 years." Damn, we can't count this one either because it's also just echoing the same original study.

No worries, plenty of independent sources left, including detailed reports from the World Health Organization, the Global Alliance for Vaccines and Immunization (GAVI), the Vaccine Impact Modelling Consortium (VIMC), and the Bill & Melinda Gates Foundation — all of which confirm that vaccines have saved 154-million lives. Oops, we can't really count any of these sources because their reports —all of them— are also based only on the exact same study. Plus, these groups *funded* the original study, so they could be slightly conflicted. Also, three of these groups were established by the Bill & Melinda Gates Foundation, and all of these groups are funded by the Bill & Melinda Gates Foundation.

Okay okay, maybe we have to accept that there's actually only one source for this widely-known fact about the 154-million lives saved. And I guess we also have to accept that the only study to ever reach this conclusion was funded by the very groups that promoted the conclusion, and benefit from it. And maybe we have to accept that the study was funded, ultimately, by just one sponsor (the Bill & Melinda Gates Foundation) working through several entities.

But none of that means the study wasn't objective or credible. We can still be quite impressed by their methods:

> "We developed a standardised analytical framework to estimate vaccine impact per fully vaccinated person over time, synthesising the results of 22 models and applying ***regression-based imputation*** methods to ensure geographical and temporal completeness."

Regression-based imputation is when you fill in missing data with assumptions and estimates, and then develop theories about the relationship between variables. It's quite understandable there'd be plenty of missing data and many variables in a model being applied to the whole planet over a 50-year period.

They didn't stop at regression-based imputation, by the way. They also "synthesised age-specific vaccine coverage estimates from four data sources: WHO Immunization dashboard; WHO Supplementary Immunization Activities Database; WHO Polio Information System; and the Vaccine Impact Modelling Consortium."

If we put aside for the moment that all four of those data sources also happened to have funded the study, we can focus instead on the quality of their process: When years of key data was unavailable, the modelers "linearly extrapolated" from later years. Then they applied their estimate "to an anchored 0% coverage in 1974." As one does.

Far from just creating a collection of guesses and assumptions, "impact estimates were derived directly through simulation of published transmission models for measles and poliomyelitis." In other words (and we need some other words), they based their simulations on past models. As one does.

But they didn't stop there. To accurately ascertain how many lives were saved by the polio vaccine, for example, they "ran novel simulations of an existing dynamic model," meaning they used a new type of simulation on an old model. They also developed "novel approaches to synthesising diverse sources of model estimates, accounting for non-linearity in vaccine impact, and extrapolating model outputs to locations without such estimates." I can't tell if that sentence is self-explanatory, self-evident, or stating the obvious, but whichever, it gets more impressive each time I read it.

To estimate how many lives were saved by just the measles vaccine on its own, they used two different dynamic models from the past — not one, but two — and estimated how many "years of full health" were gained as the result of vaccination. Apparently, the modelers believe that measles vaccine adds years of "full health," not just inoculation against measles infection.

Measles offers an interesting case study, given that the number of American children who died from measles was zero starting in 2000, and was zero again every single year for the next two decades, and then zero again in 2021, 2022, 2023, and 2024. Reality creates a particularly thorny statistical problem when you're building a model to show how many lives were saved by the measles vaccine: The problem is that the measles death rate for all those years applies to the ***millions of American children who were <u>not</u> vaccinated against measles*** as well as the millions who were vaccinated. Both groups have the exact same death rate: zero. [1] [2]

Unlike models, this inconvenient fact is not derived from estimating or guessing, extrapolating, assuming, or imputation. It's the kind of fact the esteemed modelers at Imperial College have to set aside in order to show the world that vaccines saved 154-million lives. They have the authority to set aside pesky facts because Imperial College models are the global gold-standard for guesswork, and anyone who questions their reputation need only look at their record:

---

1  2 

- 2002: Imperial College modelers predicted **150,000** deaths in the UK from Mad Cow Disease [1]
  Actual number of deaths: **177**
- 2005: They predicted 200 million could die from bird flu [2]
  Actual number of deaths: **282** (over 6 years)
- 2009: They predicted 65,000 swine flu deaths in the UK [3]
  Actual number of deaths: **45**
- 2020: They predicted up to 179,000 Covid deaths in Taiwan in the first full year of the pandemic [4]
  Actual number of deaths: **10**
- 2021: The modelers predicted 5,000 deaths per day from Omicron [5] [6]
  Actual number: **300**

Okay, maybe don't look at their record, and instead look at the star epidemiologist who runs the modeling team, Professor Neil Ferguson. Despite (or perhaps because of) the dramatic predictive failures listed above, Ferguson was tapped to serve on the British Government's Scientific Advisory Group for Emergencies (SAGE). It was his models that were used to justify closing schools, banning public events, shuttering businesses, canceling church services, social distancing, forcing isolation, and lockdowns. Ferguson's intense advocacy for extreme strategies of "suppression" (his word) earned him the nickname "Professor Lockdown."

Turned out the nickname didn't suit him well, because during the period of the Government's strictest enforcement of social distancing and imposed isolation at home, Professor Lockdown violated his own rules in every way a person could. He invited his lover (a married woman living in another household) to come over to his place more than once, including when Ferguson himself had recently tested positive for Covid.

Though the public was so unforgiving that Ferguson had to resign from his prestigious Government position, his affair and hypocritical rule-breaking are of little interest to us here. Rather, I'm focusing on that record of wildly far-off predictions:

- 150,000 predicted deaths from Mad Cow Disease — off by 847X (847 times higher than reality)
- 200-million predicted bird flu deaths — off by 700,000X

- 65,000 predicted swine flu deaths — off by 1,444X
- 179,000 predicted 2020 Covid deaths in Taiwan — off by nearly 18,000X

Never deterred by reality, Ferguson and his team next undertook a study to prove that their own lockdown recommendations had been successful beyond measure — which apparently, they could measure. Because of their advice to governments, they concluded, 3-million lives were saved in 11 European countries. After thus grading their own homework and awarding themselves that gold star A+, they next set out to show how many lives were saved worldwide by the Covid vaccines. Turns out it was a big number: 20-million people.

All kinds of criticism was lobbed at Ferguson for that claim, including overreliance on speculative modeling, overestimation of vaccine impact, oversimplified assumptions, lack of transparency about data gaps and methodology, generalizing about countries and regions that are quite different, his history of failed models, ignoring the impact of natural immunity, bias from funders (the Bill & Melinda Gates Foundation again), and oh yes, neglecting adverse events from the vaccines themselves, including the most adverse event of all: death.

Speaking of ignoring vaccine-related risks and deaths, in that report claiming 154-million people would have died if they hadn't been vaccinated, did vaccine-related deaths merit even a word or two?

No, not even a word or two. Literally. Here's a sampling of words that never appear in their 7,000-word report about the history of mass vaccination:

adverse event	side-effect	injury	harm
reaction	autism	myocarditis	brain damage
unwanted effects	inflammation	vaccine-induced poliomyelitis	
seizure	blood clot	neurological	Simian Virus 40
auto-immune	heart	heart failure	neuropathy
cardiac arrest	allergy	transverse myelitis	fatality
Guillain-Barré	convulsions	Bell's Palsy	stroke

You get the idea.

A quick note about ambitious and complex models like those created by the Imperial College: They rely upon many parameters, and even a small change to a single parameter can create exponential differences in the outcome.

For example, by assuming an erroneously high death rate from measles, the study claims 94-million deaths were averted just by measles vaccines. If the estimated fatality rate from measles is lowered to reflect reality, or if they account for factors like improvements in sanitation, healthcare, and nutrition, the estimated deaths averted will drop by millions.

If the study assumptions were adjusted to more proportionately recognize death rates from measles in China, Russia, the US, and every European country (effectively zero in all those countries), that would dramatically shrink the global estimate of lives supposedly saved.

The data on measles deaths demonstrate that even without universal vaccination (which the world doesn't have), measles mortality is negligible in many countries, and measles deaths happen mostly in places with poor nutrition and sanitation — not just countries with low vaccine rates. Again and again, reality undermines the study's assumptions. But never mind, that sole original study promotes the grandiose estimate that ***5.7 billion*** years of "full health" have been gained as the result of just the measles vaccine alone — and they've created a similar number for every vaccine.

Take tetanus, for example. In the United States, ***over a 10-year period***, there were 13 tetanus deaths, all in elderly people. In 2021, 42 of the 44 countries in Europe had zero tetanus deaths. That year, there were zero tetanus deaths in Australia, New Zealand, and Canada — and one tetanus death in Russia. Looking at tetanus deaths by population, there were effectively zero tetanus deaths per 100,000 people in all 19 countries of South America and Central America ***combined***. Same for China, Japan, South Korea, Singapore, and all countries in Europe ***combined***. In fact, fewer than 10 countries on Earth have a tetanus death rate above 1 per 100,000 people.

Whatever the global death rate from tetanus actually is, ***more than 1.5 billion Earthlings are not vaccinated against tetanus***. And tetanus deaths are vanishingly rare for them too.

For a child to be considered fully vaccinated against tetanus requires five injections, the first at 2-months old, another at 4-months, another at 6-months, another around 15-months, and the final/fifth dose sometime before 6-years old. That's what a parent has to do in order to maybe get their child the miniscule beneficial difference between being vaccinated and not being vaccinated against tetanus. What's the risk of tetanus in the US? Well, to die of tetanus an American child has to sustain a deep puncture wound that happens to be contaminated with the tetanus bacteria (which is almost unheard of), then leave that wound unwashed and untreated, and then be the one in 12-million Americans who actually contracts tetanus each year, and then be the *one in every 165-million Americans* who dies of tetanus each year.

One-in-165-million is pretty exceptional odds already, but parents who want to improve those odds when it comes to tetanus might also want to consider a few of the ***rare-but-real*** adverse events associated with those five doses of the DPT vaccine:

Seizures	High Fever
Vomiting	Diarrhea
Prolonged convulsions	Progressive neurologic disorders
Encephalopathy	Lowered consciousness
Unresponsiveness	Coma

It's reasonable to wonder if those adverse events happen more often than once in 165-million Americans, but the Imperial College report sets such questions aside in order to make the dramatic claim that ***1.4 billion*** years of full health have been gained as the result of the tetanus vaccine. As you look at this world map, doesn't it make sense to consider giving children the series of 5 tetanus injections only in those countries where it might matter?

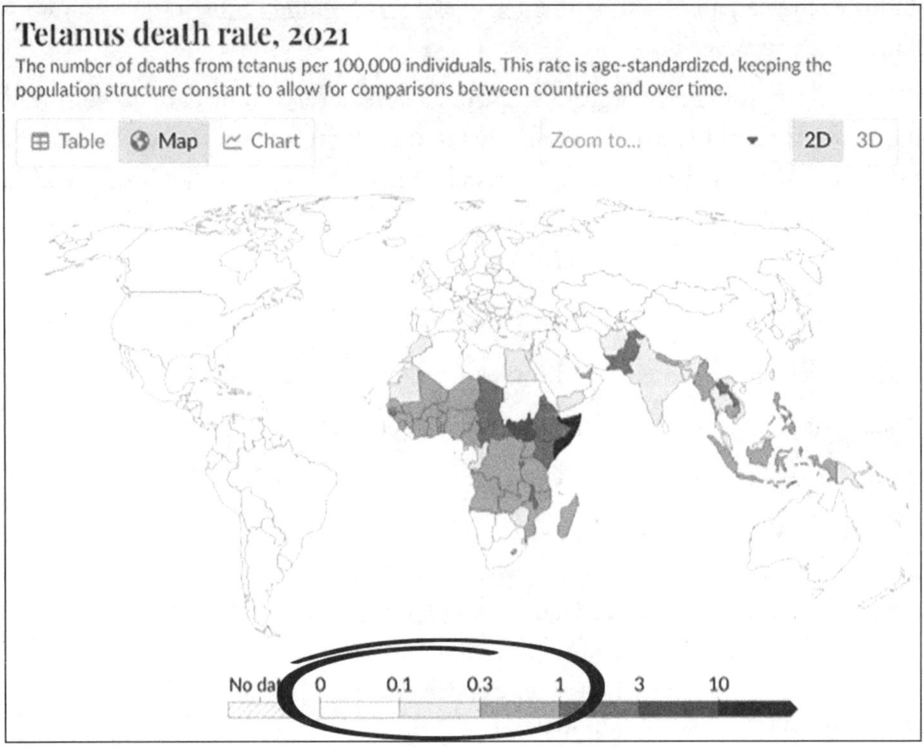

Tetanus death rate, 2021

The number of deaths from tetanus per 100,000 individuals. This rate is age-standardized, keeping the population structure constant to allow for comparisons between countries and over time.

Table Map Chart Zoom to... 2D 3D

No data 0 0.1 0.3 1 3 10

There's a QR code below linking to the underlying report.[1] People who dig into it can choose between two broad possibilities:

1. 154-million lives saved is a headline-grabbing claim, bought and paid for (and amplified) by biased stakeholders in order to affirm, encourage and expand mass vaccination. In other words, the claim is promotion, not science.

OR...

2. The number is accurate and verifiable, discovered by an unbiased, unconflicted group of geniuses.

1

CHAPTER THIRTEEN (10 MINS)

"Safe and Effective"

With all the good reasons to distrust Big Pharma companies, why are they so often afforded the kind of trust previously reserved for companies that *didn't* commit lethal criminal conspiratory fraud against their fellow citizens?

The question is even harder to answer after we take a fast journey through some intentionally forgotten Pharma history: After FDA approved Vioxx, for example, there were many litigations related to the inconvenient fact that the drug doubled the risk of heart attack. How many litigations? Oh... 27,000 litigations, but who's counting?

Merck was criminally fined almost a billion dollars for overstating the drug's safety with a now familiar refrain: "safe and effective."

Much like these days, when CDC and FDA are receiving and ignoring millions of reports about adverse reactions to Covid vaccines (e.g. myocarditis, stroke, blood clots, death), Merck told jury after jury that heart attack deaths had nothing whatsoever to do with their wonder-drug. Merck fought each lawsuit like... well, like a Pharma company fights lawsuits, often accusing their injured customers of falsifying data.

Pot, kettle.

When a jury awarded one widow $253 million, Merck appealed, and the award was overturned. A bunch of other lawsuits followed, with Merck winning some, losing some — until a class-action lawsuit concluded that Merck had violated the law by selling a drug that was unsafe, because of, you know, doubling the risk of heart attack or some such thing. And then...

Merck agreed to a mass settlement of ***$4.85 billion*** to end thousands of individual lawsuits. And then...

Merck announced a settlement with the US Attorney's Office, resolving $950 million in criminal fines that had been levied against the company.

Did that end it? Nope, litigation with several states is still ongoing. But the real punchline is...

Vioxx is returning to market. Yes. We can look forward to breathless news reports about another (likely renamed) wonder-drug that's safe and effective.

The story of Vioxx is not a cautionary tale; rather, it's a long-running series acknowledged by a group of concerned scientists in 2017:

> *To increase the likelihood of FDA approval for Vioxx, the pharmaceutical giant Merck used flawed methodologies biased toward predetermined results to exaggerate the drug's positive effects.*

You mean like nearly every other Pharma company does with its products?

> *Merck's manipulation also included a pattern of ghostwriting scientific articles. Internal documents reveal that in 16 of 20 papers reporting on clinical trials of Vioxx,* ***a Merck employee was initially listed as the lead author*** *of the first draft.*

You mean like nearly every other Pharma company does with its products? [1]

There's more than enough blame to go around, because it took literally years before Vioxx was pulled from the market (by Merck itself, not by FDA). That's hardly FDA's only failure. A more recent example is far worse: New and insufficiently-tested mRNA products were injected into billions of people, including children and even infants. Then FDA tried hard to keep the clinical trial data secret from the public. To be more accurate, FDA was perfectly willing and happy to release all the information to the public, but only if they were allowed ***55-years*** to do it. To be even more accurate, FDA (along with Pfizer) petitioned the Court together, asking for ***75-years*** to disclose all the information. [2]

Does the fact that FDA fought so hard to keep the safety results secret from the public make you curious?

Curious perhaps — but no reason to be surprised, since FDA has typically been run by people like Robert Califf, who also presided over approval of high-dose hydrocodone drugs that were later recalled. Before and between his two tours of duty at FDA, Califf was an executive with three drug companies, and also a

Robert Califf

1  2 

paid consultant to the 15 biggest Pharma companies. And finally, he was a cheerleader for Vioxx.

So of course he was Commissioner of FDA. (And of course Pharma insiders Scott Gottlieb, Lester Crawford, Mark McClellan, David Kessler, Frank Young, and Margaret Hamburg were also commissioners of FDA.)

Should we have expected more skepticism of Pharma from CDC, recently run by Dr. Rochelle Walensky? No. Here she is displaying her willful gullibility about Pfizer's marketing claims:

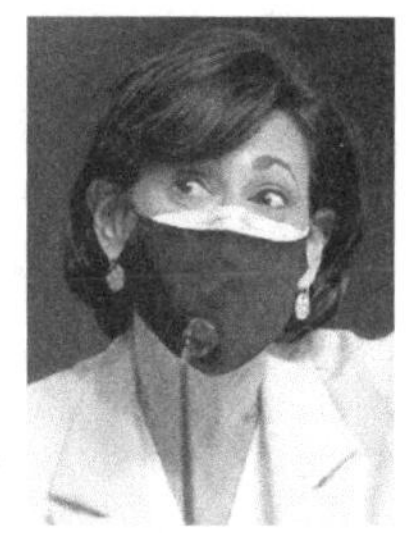
Rochelle Walensky

> "I can tell you where I was when the CNN feed came that it was 95 percent effective, the vaccine. So many of us wanted to be hopeful, so many of us wanted to say, okay, this is our ticket out, right, now we're done. So I think we had perhaps ***too little caution and too much optimism*** for some good things that came our way."
>
> "Nobody said waning, when you know, oh this vaccine's going to work. Oh, well maybe it'll wear [laughs], maybe it'll wear off."
>
> "Nobody said what if [it's] not as potent against the next variant." [1]

She couldn't predict that vaccines might be less effective against new variants? You mean, like every flu vaccine has been every year for decades?

She couldn't predict that the vaccine would wane, like the flu vaccine has done every year for decades, like the measles vaccine and pertussis vaccine and tetanus vaccine?

Dr. Walensky's catchphrase, "too little caution and too much optimism," could be a lyric in CDC's most famous song, *Safe and Effective*. To better understand how the agency always concludes that every vaccine is safe and effective, we can consider a vaccine so old that all its risk factors are well-known: the smallpox vaccine.

From the CDC website:

> ***The smallpox vaccine is safe***, *and it is effective at preventing smallpox disease.*

Let's see what *safe* means to the CDC, in their own words:

> *Serious Side Effects of Smallpox Vaccine*
>
> - *Heart problems*

---

1 

- *Swelling of the brain or spinal cord*
- *Severe skin diseases*
- *Spreading the virus to other parts of the body* ***or to another person*** *[huh?]*
- *Severe allergic reaction after vaccination*
- *Accidental infection of the eye (which can lead to blindness)*

The risks for serious smallpox vaccine side effects are greater for:

- *People with any three of the following risk factors for heart disease: high blood pressure, high cholesterol, diabetes*
- *People with heart or blood vessel problems, including angina, previous heart attack or other cardiac problems*
- *People with skin problems, such as eczema*
- *Women who are pregnant or breastfeeding*

That predicts a giant number of people risking brain swelling, heart problems and blindness, given that millions of Americans have had heart attacks "or other cardiac problems," 17-million Americans are pregnant or breastfeeding, 31-million have eczema, 34-million have diabetes, 76-million have high cholesterol, and more than 100-million have high blood pressure.

Describing the people most at risk of serious side effects from the "safe" smallpox vaccine, CDC included people with a "family history of heart problems."

Do you know anyone who <u>*doesn't*</u> fit that category?

So while CDC definitively stated "***The smallpox vaccine is safe***," they then listed people for whom it's not safe — and that just happens to be, oh look, the majority of Americans.

But Gavin, since there's no outbreak of smallpox, why worry about what was on CDC's website? Because... wait for it... CDC actively promoted that same dangerous smallpox vaccine for people worried about monkeypox — and nearly a million Americans have had the pleasure. (I guess it's people with no family history of heart problems.)

And why have just one smallpox vaccine when you can have two?

NATION WORLD

CDC: 1 dose of vaccine protects against monkeypox, 2nd dose encouraged

—

Experts urge a second dose for full protection, but people who received a single dose of the monkeypox vaccine appeared to be significantly less likely to get sick. [1]

You now know that when CDC said "safe and effective," it's because CDC never met a vaccine they didn't like. It's a belief system — not science.

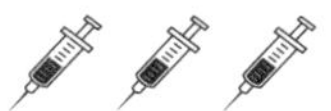

Because most people don't have time to take a deep dive into the safety of childhood vaccines, they must choose who to rely upon for accurate information. Should we trust vaccine-makers to tell us the truth? Can we trust public health officials or the professional medical associations that parrot what they hear from public health officials? All these groups would say American citizens are not qualified to conduct their own research into The Science. I suggest that making good parental decisions derives from common sense, not science.

For example, how would we feel if every one of the top vaccine manufacturers has been criminally fined for the worst sorts of offenses?

Let's find out how we'd feel...

Pfizer: For illegal marketing of its products, the company was forced to pay the largest criminal fine in US history. Plus $1 billion more in a 2009 settlement to resolve violations of the False Claims Act. I almost forgot another $240 million in criminal fines, and another $190 million in 2004 to resolve false claims made to Medicare and Medicaid. That's all — other than $60 million the same year for kickbacks and off-label marketing, and $40 million in settlements with whistleblowers. And just one more: $75 million for fraudulent marketing. Sorry, another small one, hardly worth mentioning: $15 million for paying kickbacks to healthcare providers.

---

1 

Johnson & Johnson: This particularly trusted maker of vaccines given to children paid about $5 billion to multiple states and local governments for its role in the opioid crisis. You can decide if any of these actions lessen your trust in the company: deceptive marketing, downplaying the risks and overstating the benefits of their products, false claims made to mislead doctors, patients, and regulators, tricky promotion to get doctors to prescribe their products more widely, producing and distributing products they knew to be dangerous, producing and distributing Fentanyl products, and failing to adequately warn about the risks of their products.

Earlier in the book, we already covered the billions J&J paid (4.7 billion so far) in connection with baby powder that they knew for decades had cancer-causing ingredients that harmed people, so I'll skip that and note another $2.2 billion in penalties for illegal marketing and kickbacks to doctors and pharmacies. Then there's the $5 billion they paid in 2013 for selling a product they knew was dangerously defective. I'd like to stop here, but Johnson & Johnson's harmful behavior didn't stop; they paid another $1.1 billion in penalties for failing to warn about the risk of heart attack and strokes from one of their products. Should I even bother to mention the $33 million for their false and misleading marketing that overstated benefits and downplayed risks of Depo-Provera? There's also the $70 million for bribing doctors and pharmacies to prescribe drugs that were not appropriate for the conditions being treated.

Is that so bad?

GlaxoSmithKline: Not to be outdone by Pfizer or J&J, these good corporate citizens paid $3 billion in penalties for fraud and failure to report safety data. What kind of information did they conceal? The usual: increased risk of heart failure. Glaxo apologized and moved on.

In 2014, the European CDC reported that Glaxo released gallons of live polio virus solution into a water-treatment plant in Belgium. The concentrated virus made its way into three rivers in Belgium. It had already been a bad year for the company, having been found guilty of systematically bribing doctors, hospitals and government officials to boost drug sales, through a network of 700 middlemen and payments estimated at $150 million. GSK initially called the allegations a "smear campaign," but their good conscience eventually led them to admit the illegal activities. Glaxo apologized and moved on.

Merck: I've already written about Vioxx, and I don't want to pile on, so I'll ignore Merck's $650 million criminal fine for paying kickbacks and overbilling Medicare, and instead recall that time when two of Merck's virologists became whistleblowers and revealed that the company had tested their flagship product, the MMR vaccine, against a cherry-picked variant of the mumps virus rather than against the actual virus children might encounter. And guess what: The MMR tests came out really good — until it was revealed Merck had added rabbit blood to human samples in order to boost favorable results. Hey, nobody's perfect.

Eli Lilly: This slightly better corporate citizen enjoyed a criminal fine of $515 million in 2009, as well as a civil settlement that could reach $800 million, for a total of $1.4 billion, all resulting from their product Zyprexa. A couple of years earlier, they had to pay $1.2 billion to settle about 32,000 cases with customers who were injured by the same product, plus $47 million paid to a bunch of states for improper marketing.

Why were people so pissed off at Eli Lilly? Was it that Zyprexa caused swelling of the hands, feet, arms, legs and face? A little bit. Was it the problems swallowing or the drooling or the twisting body movements? Sure, in part. Was it the rapid weight gain and diabetes? No, those were the lucky customers.

Was it that Zyprexa caused stroke and heart failure and sudden cardiac death? Probably explains some of the resentment — but most of all, it was that Eli Lilly knew about these side effects and kept on marketing and selling the product.

Putting morality aside, it turns out that Eli Lilly made great business decisions because despite all those massive penalties and fines and settlements, customers still went crazy for Zyprexa — so crazy that the drug produced $40 billion in revenues.

I could share similarly rosy accounts regarding Sanofi, Novartis, and Roche — but all we'd see is more price gouging, failure to disclose side effects, bribery, corruption, defective products, delayed recalls, lies, misuse of political influence, and cheating on clinical trials.

In the 31 years before 2021, Pharma manufacturers paid $62.3 billion in penalties, a number so large that you might wonder how they survive as an industry. Well, they don't survive — they thrive. The penalties amount to a small percentage of the *trillions* made by the largest drug companies during those same years.

By the way, if you're hoping all those penalties helped Big Pharma to find the Lord, sorry to disappoint. Most are repeat offenders, with Pfizer leading the pack when it comes to the sheer number of cases. [1]

1 

This probably isn't a good time to ask if Pharma has earned your trust when they insist that their products are safe to inject into children. Remarkably, despite all that we all know about the deceit and treachery of these truly criminal enterprises, many of us chose to believe that somehow, vaccines are the exception, the one product Pharma uncharacteristically manufactures with great care. Vaccines are not an exception. In fact, given that vaccine-makers can't be sued no matter how toxic or contaminated or injurious their products are, can anyone reasonably believe that pharma companies concoct and test and manufacture and market and sell these products with the greatest care and honesty? Take a moment and see if you can think of anything, anything in their past behavior or current incentives that would support such a belief.

Let's get back to where we started, back to how the Institute of Medicine debunked any possible vaccine-autism link. For a task so important to the well-being of America's children, you might imagine IOM experts examined patients, studied blood tests, reviewed medical records, read autopsy reports, scrutinized tissue samples, conducted experiments, tested various compounds on animals, or mixed chemicals in beakers to see how they react. No, nothing at all like that.

The IOM committee reached their rock-solid conclusions without ever donning white lab coats or examining an injured child. Instead, their work was done in closed-door meetings that they were confident would never become public. As it turns out, however, the brilliant experts failed to predict that someone with a conscience might leak the transcripts of the discussions. And someone did. And those transcripts are about to give you a direct look into exactly how the sausage is made when the US Government decides something needs to be debunked.

Speaking of sausage, imagine you're touring an actual sausage factory. The escort hands you a clean white smock to put over your clothes. Employees occasionally ask you to keep your distance from some macabre vat or greasy conveyor belt. Stand back a bit, they warn, sometimes suggesting you stand back even farther. Ten feet away from a big grinding mechanism, there's a bold red line painted on the floor, and your guide admonishes you, "Stand behind the splatter line."

That phrase —the splatter line— gets your attention. When the grisly machines start up, you're a bit alarmed at what you see and hear, but more than anything you're grateful someone painted that red line on the concrete floor, and grateful it's a good distance from the carnage.

How is the sausage made when the Government and its industry partners decide they had better debunk some concern of great consequence, like Agent Orange and injuries from pharmaceutical products? Or Gulf War Syndrome, cancer-causing baby powder, mercury in childhood vaccines, poison baby formula, toxic baby food, burn pits, SIDS and anthrax vaccines? Or Reye's Syndrome, cardiac injuries from mRNA shots, and autism? You're about to see firsthand.

Stand behind the splatter line.

CHAPTER FOURTEEN (9 MINS)

Let the Word-Games Begin

Before you meet the medical wizards of Oz who served on IOM's Immunization Safety Review Committee (the wizards who debunked any possible link between any vaccine and any child's autism), note that the Institute of Medicine is not a government agency or department; it's a totally private organization often hired and funded by the Government — and also hired and funded by private industry, including generous companies like Johnson & Johnson, Merck, Pfizer, AstraZeneca, Eli Lilly — you get the idea

The head of the IOM is Victor Dzau, who is paid a million+ dollars a year. One of his initiatives is to get even more funding from private companies. So that's a thing.[1] [2]

Victor Dzau

The IOM is part of the National Academy of Sciences (NAS), which sounds like a government entity, but isn't. It too is a private organization — and thus allowed to conceal how much money the IOM was paid to debunk the vaccine-autism link, Agent Orange, burn pits, SIDS, and everything else.

Before we eavesdrop on the deliberations of the IOM committee that forever debunked any vaccine-autism link, here's a fast look at a more recent National Academy of Sciences committee, this one tasked to apply its expert scrutiny to why, every year, US taxpayers pay billions for expensive pharmaceuticals that are simply thrown away. That happens because pharma companies intentionally ship many of their drug products in oversized vials that, once opened, usually cannot be resealed or saved for other patients. Yet pharma gets paid for every grain of powder and every drop of secret potion, even that which has to be discarded because of the tricky packaging scam.

---

1 2

When Congress needed some independent, objective advice on the matter, that's exactly what was being sold by the National Academy of Sciences — the words are right on their homepage: "independent, objective advice."

But wait, there's more: They also "confront challenging issues for the benefit of society," and provide "trustworthy advice." And they have core values, once more invoking independence and objectivity. But that's not all; customers also get "Rigor, Integrity, Inclusivity, Truth." When Congress needed help to understand and overturn pharma scams, the National Academy of Sciences probably sounded like a bargain.

Too bad it didn't work out that way.

The NAS committee report turned out to be decidedly pro-pharma, discouraging the Government from trying to recoup any money from pharma companies, and encouraging Medicare to stop tracking the cost of drug waste altogether. It was a great report for Pharma — not so great for the taxpayers.

The report was missing a few interesting details that came out later; for example the fact that one committee member, Dr. Kavita Patel, was said to have earned *$1.4 million* serving on the board of a pharmaceutical corporation. She had also just joined the board at biotech company, Sigalon, reportedly earning the good doctor another few hundred thousand dollars. By the way, Sigalon warned its investors that government "cost containment measures could significantly decrease the price we might establish for our products," meaning this particular committee might pose a ***risk to the company's bottom line.*** [1]

Dr. Patel is a woman of many trades (medical doctor, venture capital investor, venture capital advisor, venture partner, serial board member, media contributor,) who adores pharma companies and pharma products, even those she knows little about, like the ones that are brand new. When the FDA released its first promotional reports on Pfizer's mRNA vaccine, Dr. Patel was on TV doing one or more of her jobs.

How did she feel about the new Pfizer vaccine? "Gleeful," she told MSNBC, with shoulder-shaking giggles of joy.

---

1 

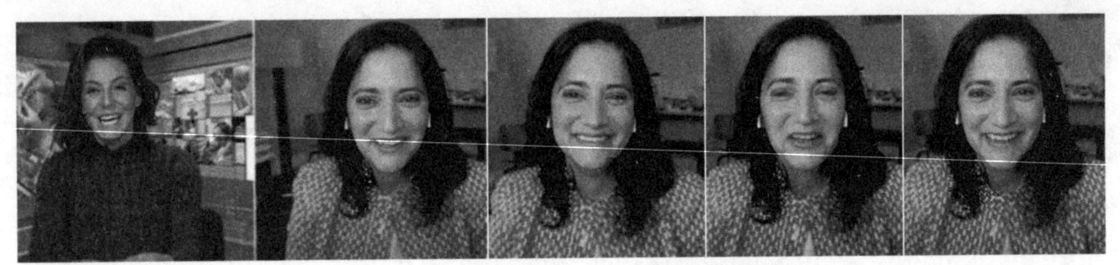

MSNBC Interviewer Kavita Patel

"Of the people who received the vaccine," she celebrated, "they are 95% less likely to get mild, moderate or severe Covid compared to a group of people that don't get the vaccine."

"That's the right way to interpret that," she added, while offering the wrong way to interpret that. (It wasn't true.)

Best of all, she gleefully exclaimed, there were "certainly no serious adverse effects, *to speak of*."

In fact, there were many, many serious adverse effects from the Pfizer mRNA vaccines — but I agree there were none Dr. Patel was willing "to speak of."

When the giggling subsided, she said "I want to warn Americans," as if about to impart something important, but all she said is that some Americans might have very mild arm ache from the injections. Not so much warning about cardiac injury or being more likely to get Covid again and again, or dramatic impact on menstruation and sperm counts, or strokes and blood clotting — and not a word of warning about that really pesky adverse effect: sudden death. [1]

Another committee member had received consulting income from more than ten pharmaceutical companies, eight of which had earned millions by billing Medicare for wasted drugs.

The conflicts were so obvious that NAS had to cough out a statement acknowledging that "for-profit organizations with a direct financial interest in the outcome of a study" should not be involved in funding the study, "***except in rare circumstances***."

By the way, there were similar questions about NAS in 2014 for failing to disclose conflicts among committee members who advised federal officials on opioid

---

1

use, and again in 2017 when opining on genetically modified crops, and again in 2018, when NAS acknowledged more than $5-million in gifts from Merck.

One of the NAS committee members joined a group of people who worked for the trade group that represents major pharma companies (phRMA), and they together wrote an article arguing that medications should be valued *not for their actual benefit*, but rather for the innovation potential that might arise from making new therapies. Could you give me that again? The actual benefit from a given drug should be secondary to the possibility that maybe perhaps potentially someday, this drug that isn't all that beneficial might lead to new therapies. Enough about conflicts for NAS committee members; how about conflicts for the NAS itself, given their responsibility to select and oversee people who will advise the Government on waste and improper billing by pharma companies? Uh oh, among NAS's generous donors were companies with millions of dollars at stake in the findings of this very committee. [1]

That brings us back to IOM. Though congratulating itself for being "objective and credible," those might not be the exact words that come to mind as you read their actual conversations. The people you're about to meet are treated like revered deities of the medical cartel, experts who graciously accept awards they give each other at dinners we all pay for. They are among the excuse-makers and concealers of public health mistakes and lies, the medical mandarins who pull the levers of life and death as if playing slot machines.

The presumed work of this particular IOM committee was to determine whether any childhood vaccine could harm any child, including whether or not any of these products might be linked to autism in any children. Though they knew everything you've read up to this page, the committee members served as dedicated decoys pointing away from the truth.

See if I've overstated things...

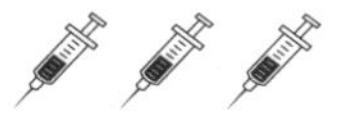

As we enter the "CLOSED SESSION" on January 12, 2001, no scientific or medical expertise is needed to understand the two preambles recited by Study Director Kathleen Stratton. First was her optimistic pronouncement that "closed

---

1 

session transcripts will never be shared with anybody outside the committee and the staff." While that turned out to be wrong, her second pronouncement was dead-on accurate:

> "The point of no return, ***the line we will not cross*** in public policy is '*Pull the vaccine, Change the schedule.*' We wouldn't say '*Compensate [the injured].*' We wouldn't say '*Pull the vaccine.*' We wouldn't say '*Stop the program.*'"

Kathleen Stratton

The Chair of the committee, Dr. Marie McCormick, further cemented the conclusion that the co-opted committee had already committed to reach before they even started:

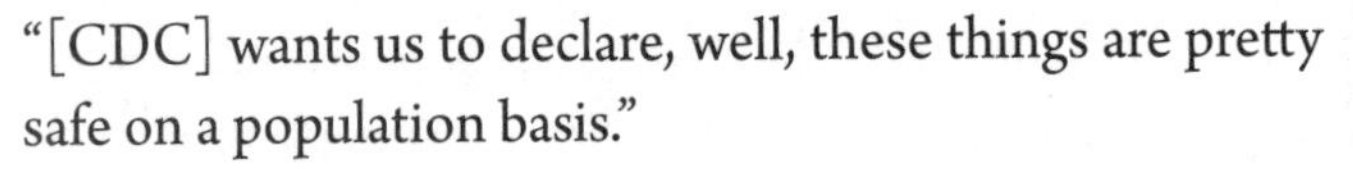

> "[CDC] wants us to declare, well, these things are pretty safe on a population basis."
>
> "We are not ever going to come down that [autism] is a true side effect."

Marie McCormick

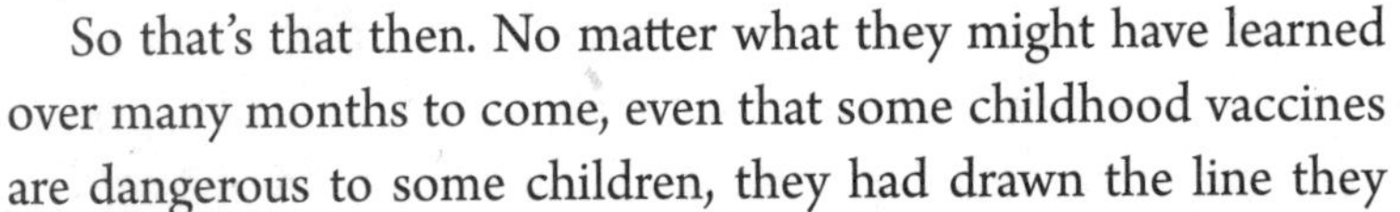

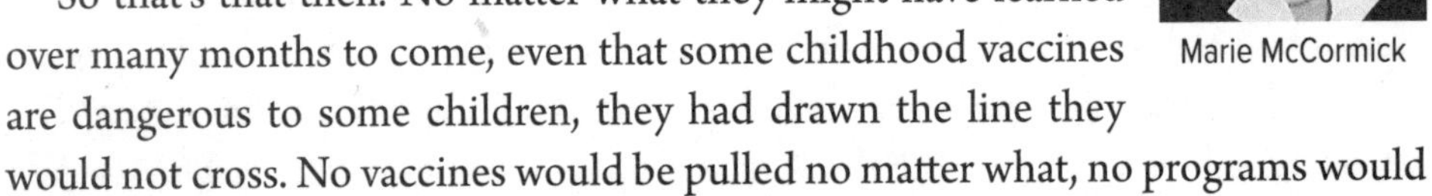

So that's that then. No matter what they might have learned over many months to come, even that some childhood vaccines are dangerous to some children, they had drawn the line they would not cross. No vaccines would be pulled no matter what, no programs would be stopped no matter what, and no schedule of childhood vaccines would be changed — no matter what. With the ending embedded in the beginning, they marched into their real work: how to artfully say what they had to say, while artfully avoiding what they must not say.

> Dr. Berg: I don't know how long it will take us to figure out what the question is. I am a veteran of one panel that took six days for a group about this large to figure out what the question was… it can be a formidable issue. I don't know what the question is, whether it is MMR or whether it is measles [vaccine].

Albert Berg

> Dr. McCormick: The question is MMR.
>
> Dr. Wilson: Are we to look at thimerosal?
>
> Dr. McCormick: Not this round.
>
> Dr. Wilson: Wait a second. Not this round? But if we are going to look at autism and we have three candidates, can we really fundamentally look at them in isolation? In fact,

Christopher Wilson

> in the real world, they don't occur in isolation. Individuals that got MMR vaccine also received vaccines with thimerosal.

Right he was, but then Dr. Berg interrupted to ask more questions about the question.

> Dr. Berg: Excuse me… we're going to have to have a method for how we focus the question many times. This is one of the questions we need to focus on… how are we going to form the question… in general terms, what process are we going to use to focus the question? How are we going to discuss it? I would like some reassurances about the general processes before we get to the specifics of autism and MMR.

Dr. McCormick assured Dr. Berg that she planned to dedicate a whole day to his concerns about the question, but Dr. Berg was stuck: "The issue of specifying the question is an important step. I would like to know how this panel is going to specify the question."

> Dr. Wilson: Why don't you make a proposal?

> Dr. Berg: Do we look at just burden of suffering? Do we look at squeaky wheels? What information will this panel collect in order to decide what is the question that we are going to address?

Michael Kaback

As if speaking for us all, Dr. Kaback said, "I am not sure what you just said. Which information do you collect in order to determine which question you address, or is it the other way around?"

> Dr. Berg: Yes.

"We Can Parse it Any Way We Want"

Right from the start of the closed-door meetings, Dr. Goodman was always willing to confidently lay out his position in a way that made it impossible to understand:

> "In the end, the bottom line is, the only verdict that is of importance is whether we say a causal relationship [between vaccines and autism] is ***Likely, Suggested, Unlikely, Inadequate***. That is the only metric. As soon as we start introducing other words that… sort of sidestep to causality, that we say represents ***Intermediate Levels of Causality — but we are not going to say causality*** — I think we introduce confusion."

Introducing confusion, you'll soon see, is the specialty and purpose of the experts who worked behind closed doors and closed minds to avoid answering America's pressing question: Might there be any link between any vaccine and any neurological injury to any child?

Dr. Goodman continued:

> "It is still a judgement about, on the basis of observational evidence, how likely it is to be causal. We can decide to subdivide the ***Suggestive*** category into ***High Suggestive*** and ***Low Suggestive*** —I don't think we should— but I don't think we should start using words like association which first of all, is a highly technical term guaranteed to be misunderstood, if you are using it as a non-technical term for non-causality."

Make sense? Of course not. This next part is easier to understand:

> "It actually ***hides the fact that we are making a judgement that it is not causal.***"

Those words ("we are making a judgement that it is not causal") were spoken on Day One, months before they completed their supposed review. The whole theatrical production that followed is comprised of scene after scene establishing that they had already decided the big issues. The rest of the play is about one thing only: how to say it — or more accurately, how to *not* say it.

> "That intermediate category is a very uncomfortable category, and everybody here knows that. It is a very uncomfortable category… and we can parse it any way we want."

When Dr. Goodman reminded the group they could parse it any way they want, he wasn't only stating the obvious. He was stating the mission. To drive it home, he did a few more laps around the parsing track:

> "I don't think we should put an ***Association*** category and then a ***Suggestive*** category. You know, you have ***Suggests an Association*** and ***Establishes an Association*** and ***Favors a Causal Relationship***; that is really a subdivision of the ***Suggestive Causality*** category."
>
> "If we want to subdivide it [and oh, did they ever], then we should decide on that. I think we have to use ***Causal Association*** in every category… we can say all the things you said: 'There seems to be a ***Strong Association*** which we can't explain;' 'we don't have any other explanation for it.' However we don't want to make a causal claim because we know in many observational

> studies blah, blah, blah, blah, blah, and explain why we are not going to put it in the ***Sufficient*** category, ***in spite of all the observational evidence***."

Dr. Goodman had just described what they supposedly knew from observational studies: "blah, blah, blah, blah, blah." In case five blahs didn't make it vague enough, he followed by restating his main point: "We can decide on whatever standard we are going to use."

"In Terms of the Terms"

When Dr. Goodman floated the proposal that they "split the ***Suggestive Evidence*** into ***High Suggestive*** and ***Low Suggestive***," Dr. Wilson threw cold water on the idea: "We need first to decide on the descriptors we are going to put into the categories, don't we?"

Michael Stoto

Dr. Stoto responded by touting the artful way the Government handled another chemical compound whose toxicity was debunked by IOM:

> "In the Agent Orange, the word ***Association*** really means *weak evidence of causality.*"

(Michael Stoto was an expert on the words they used for the Agent Orange committee; he led the staff on the Government's Agent Orange committee before being promoted/demoted to childhood vaccines.)

Dr. Wilson: "Why not just say that? I think we should just say it, and use the word ***causality***. In many lay people's minds ***Association*** means ***Causation***. In terms of the terms, let's just settle on what we are going to call it first, and then we are going to talk about how we are going to put it in those categories."

Dr. McCormick: "That is fine, but I don't want the word ***Association***."

Richard Johnston, Jr.

In the race to choose winning categories, Dr. Stoto favored a mixed model: "...take the top category from vaccines, the second category from Agent Orange, and then in between is the top from Agent Orange and the second from vaccines."

Voila? Not yet. Dr. Kaback was having trouble keeping track of all the parsing and subdividing: "Could somebody write this down?" Fortunately, somebody recorded it instead — which is why we have the transcripts.

> Dr. Johnston: I don't like the idea of reducing the categories, personally.

Not understanding what was being reduced, Dr. Goodman made a request that was outlandish given his audience: "Be clear."

Steven Goodman

Rather than being clear, Dr. Johnston responded with a pronouncement: "I don't think the way Agent Orange did it is the right way to go with vaccines." His distinction makes good sense, one chemical having harmed plenty of children, and the other having harmed plenty of children.

It's likely that most parents wouldn't want vaccines assessed the same way as a chemical weapon — but the experts know best.

Particularly Dr. Johnston knew best when it came to debunking, having previously chaired four IOM committees. Rather than borrowing from their Agent Orange scam, he preferred to look at how **the spin was** spun with other vaccines, because applying the process **used for totally** different vaccine products would be slightly better science than applying the process used for a bioweapon. He then jumped right to the conclusion they intended to announce many months in their future, expressed as if he were Hamlet:

> "*Inadequate to Accept or Reject. That is a statement. That is our statement about causality: 'We don't have good data that allows us to clearly say whether it is causal or not.'*"

Say nothing at all. Now, *that* is a statement.

Dr. Johnston, by the way, had attended another Government-sponsored meeting about vaccines and autism. The leaked transcripts of that more sinister meeting include these tributes to government transparency:

> "We have asked you to keep this information confidential."
>
> "Let me just reemphasize if I could the importance of trying to protect the information that we have been talking about."
>
> "Consider this embargoed information."
>
> "It would serve all of our interests best if we could continue to consider these data in a certain protected environment."
>
> "Given the sensitivity of information, we have been able to manage to keep it out of, let's say, less responsible hands."

> "We are asking people who have done a great job protecting this information up until now, to continue to do that." [1]

What was the big bad secret? Well, an official analysis had raised an unacceptable possibility:

> "***a direct causality link between thimerosal and various neurological outcomes***" [outcomes that shall remain unnamed, but you can guess]

Plenty of disturbing things had to be said at the secret meeting:

> "The data that are there, ***they won't go away***. They are going to be captured by the public..."
>
> "***Perhaps this study should not have been done at all, because the outcome of it could have, to some extent, been predicted.***"
>
> "...how we handle it from here is extremely problematic."
>
> "What we have here is people who have, for every best reason in the world, pursued a direction of research. But there is now the point at which ***the research results have to be handled***, and even if this committee decides that there is no association [between thimerosal and neurological injury], and that information gets out, the work has been done, and through Freedom of Information that will be taken by others and will be used in other ways beyond the control of this group."

1

CHAPTER FIFTEEN (14 MINS)

"If We Were a Group Working for Philip Morris, We'd Be Saying There's No Relation Between Cancer and Smoking"

Dr. Johnston described what CDC wanted them to do, and said, "I think we have got to do that in some way."

In what way did he mean? "A way that is dependent upon something that, if it is not data, it is some kind of argument, some kind of information that bears on how we might distribute our concern. Here I am talking about an emotional concern along some axis, and then you don't avoid dealing with anything other than causality in the official statement."

Translation:

> We are going to do what CDC asks, and we have to do it in a way that is not dependent upon data. We will depend upon some kind of argument, some way to show our concern. By doing this, our official statement won't technically avoid dealing with anything "***other than causality***."

Clearly, it was important to avoid the dreaded topic of causality.

> Dr. Johnston: Barbara Loe Fisher [a witness they heard from] could give you names. Mrs. Fisher said she had cases. I think she came up to say if you needed any cases to demonstrate the points, you could have them.
>
> Dr. McCormick: She was demonstrating causality. She was taken by your case series that you did — the Guillain-Barré and whatever, the tetanus [vaccine]. She was all ready to get you cases to prove causality.
>
> Dr. Wilson: Well, let's see them.

> Dr. McCormick: ***Let's not do that.*** Do you have a free weekend that you want to plod through them?

It turned out nobody wanted to commit any time at all to plod through anything. So plod they did not.

Dr. Stratton warned that if they accepted the information Barbara Loe Fisher offered to provide...

> "...all we are going to get a list of ***hundreds and hundreds and hundreds of kids who were developing normally, but they got their MMR and then they started to regress.***"

In other words, all we are going to get is exactly what we are supposed to be assessing.

> Dr. Kaback: Since 1994, Dick... has there been any significant new literature to 1ook at, other than what you guys looked at for the 1994 study?
>
> Dr. Stratton: We didn't look at this in 1994.
>
> Dr. Kaback: Not autism, I know, but you did look at MMR [the Mumps Measles Rubella vaccine]. Wasn't MMR — sure it was. ***You did find some associations*** — I don't want to use that word anymore — *Suggested Causality or Possible Causality.*
>
> Dr. Johnston: Relationships.
>
> Dr. Kaback: Relationships, right. Since the 1994 data set, there has been — will we have that literature provided to us in the next couple of weeks?
>
> Dr. Stratton: Of every safety concern with MMR?
>
> Dr. Kaback: No, we are only going to look at autism, correct?

Nobody answered Dr. Kaback because it became a discussion of how some meddlesome woman named Barbara Loe Fisher had provided a list of symptoms that was just too long.

> "You could take a textbook of pediatrics and find almost every disorder in there was in some way related to one or other of those symptoms, in which case they were all due to MMR vaccine or DPT, whatever she was claiming, MMR. There was at one point in her talk a whole array of symptoms and signs that were so non-specific, and if you really ascribed those to an adverse event, you were looking at a textbook of pediatrics."

Dr. Johnston was less sarcastic than Dr. Kaback, noting sincerely that "It would be kind of interesting to see if there are other vaccine adverse events being reported that might fundamentally pathophysiologically relate to autism."

That would indeed be "kind of interesting," and also right on the point they gathered to address. And so they steered clear of it.

Gerald Medoff

As the group steadfastly avoided the kinds of information that might be valuable, Dr. Medoff said, "I am sorry, I was just out of the room. What do we say now that we want additional neurologic complications of vaccines?"

He could have left the room for a month and he wouldn't have missed anything. When someone explained they'd been discussing the proposal of looking at all symptoms reported after vaccination because "they may, from a mechanistic point of view, give us some information," Dr. Medoff said, "I would use them with some caution. I think we really need to be careful..." Later he said "I would really urge caution in trying to make this thing too diffuse. It is just going to get into a mess. As soon as you have one [symptom], there are five others. I think if the CDC comes up with fairly specific questions that we have to deal with, I think autism and MMR is enough."

Dr. Kaback said, "The problem, Gerry, is that autism is very poorly defined." He explained that if you ask experts, "You are going to hear a spectrum."

> Dr. Medoff: Then we have trouble right off the bat.
>
> Dr. Stratton: Oh, we have trouble right off the bat.

Oh, they had trouble right off the bat. But all their troubles could be fixed with words. This process of dissecting, slicing and dicing their duty to the public wasn't science, of course — but it was the method by which any possible link between vaccines and autism would soon be officially and thoroughly debunked.

Dr. Johnston proposed a perfectly reasonable time-saver: "I can tell you right now, if our conclusion on autism and [the MMR vaccine] is what is in Cat Number Two [no evidence of association], we can say that right now and go home."

But they couldn't go home quite yet because, Dr. Johnston continued, "The CDC doesn't want that. That is not their idea of why they established this committee. In addition, the compensation program wants something else, too..." He had earlier explained that there was a need at CDC to support "the masters who determine the compensation or not." He was referring to the bureaucrats who determine which children among the thousands claiming vaccine-injury might receive compensation from the Government. (Almost none would, of course.)

Bennett Shaywitz

There was still more science to be done before they could go home. On the question of whether any vaccines might contribute to autism, Dr. Shaywitz stepped up to create the range of possible conclusions they might reach: "You have ***Yes there is a cause, No there is not, Maybe there is a cause***, and ***Maybe there is not a cause***, or ***More likely not a cause than a cause***, or ***More likely a cause***. So, you have those intermediaries. Then you have right in the middle: ***It is inadequate***."

The science continued: "Obviously, it is going to be hard to say the top or bottom, the extremes, but at least we have a ***Maybe, More likely a cause***, then ***Less likely a cause***, and ***More likely not a cause***."

To finish off his list, the natty doctor added one more category: "Or vice versa." That last category meant they could just flip his whole list upside-down if that would better serve science. After so much heavy lifting, Dr. Shaywitz was nearly satisfied: "Then hopefully that will take care of the CDC's worry that we are not having a gradation."

"You got it," Dr. Johnston announced, but added that "it still wasn't enough." Again, they couldn't go home. Dr. Goodman said, "I think this is the right number of categories. Whether we want to use these exact words we can debate."

And that's what they did for days more: debate words. Debate science? Not so much.

While he still had the floor, Dr. Goodman was gearing up to say something big. After warning his peers that it was still sort of rough, he rolled out his latest invention.

> "I call this a three-category system. That is, ***High, Intermediate*** and ***Inadequate***, with the other two categories being ***No evidence at all***, or ***Favors acceptance***... I actually think we have to use the phrase ***Favors but does not establish*** within that category. Just the phrase ***Favors*** is, again, open to misinterpretation the same way that *association* is. Then we have to discuss, do we want to use the word *establish*, which is a very high bar."

Dr. Stratton rescued them all from the high bar: "I think that actually you don't have to agonize over it. Not to prejudge your decisions over the next three years," she said while prejudging their decisions, "but I will bet you a hundred bucks you will never come up with a Category Five. It won't even cross your mind."

And right she was: The idea that any vaccine could be linked to autism — that never did cross their minds.

> *"In my 10 years of service in the US Congress, I have never seen a report so badly miss the mark. I have heard some weak arguments around Washington, and I can tell you that those in the IOM's recent report are very weak."*
>
> — Congressman Dave Weldon [1]

"An Argument in There That is Embedded Inside the Rhetoric"

Right before adjourning their first big day of scientific exploration, with all its impressive composition and prose, Dr. Goodman pointed out "an argument in there that is embedded inside the rhetoric that might be of some value to parse out and take pieces of, and make sure that we address it — in its pieces..."

That pile of Scrabble tiles apparently made perfect sense to Dr. Stratton, who promised that when they reconvened, they'd be ready to "start dissecting every single one of these ideas." They adjourned early that day, then took a much-needed break from dissecting, parsing, dividing and subdividing. After a few short months, they were ready to reconvene and get going again.

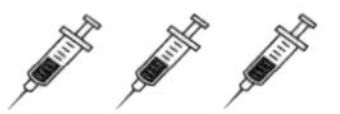

Dr. McCormick:

> "If we are willing to consider that it could occur in some other groups, even though they are not detectable at the epidemiologic level, then we are basically saying [the vaccine] can cause the adverse reaction."

And we mustn't say that, of course.

---

1 

Dr. Shaywitz said, "We can assuage the parents by saying it may happen one in a million times. We can't rule that out."

Later, Dr. Medoff continued that theme, stressing the importance of giving parents some numerical basis "for being able to say, yes I can understand that, and make a decision — the right decision — to have their kids vaccinated." To be sure parents don't make the wrong decision, Dr. Medoff proposed that IOM "give some estimate of what you think the complication rate might be, if it does occur. I think people understand that. That is giving them some numbers."

And with the numbers he proposed, the odds that started at one in a million were getting far worse:

> "If it is a problem, it probably doesn't occur more than one time in a thousand, one time in 10,000, or something like that."

Dr. Gatsonis agreed:

> "We are saying that if MMR causes this, it happens, it's such a small-scale phenomenon that it doesn't justify a change of policy."

Constantine Gatsonis

Dr. Gatsonis was able to instantly designate the risk as "small scale" because he's an expert of applied mathematics and statistical sciences. Apparently, you don't have to apply any mathematics to complete such calculations in the field of statistical science.

Dr. Medoff and Dr. Shaywitz offered ranges that differed by a hundred-fold, or a thousand-fold, "or something like that."

It was at least a brief stab at statistical science, a brief moment of acknowledging that maybe some child somewhere, maybe "one time in a thousand" might actually be harmed by a vaccine product. Other than that, the forgone conclusion was clear: The Immunization Safety Review Committee would not be reviewing the safety of immunizations.

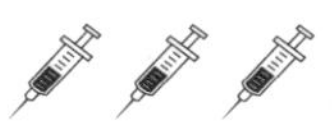

Dr. Johnston suggested they frame the questions they want to answer: "Is there an epidemic of autism, number one? Number two, can you tell from the epidemiological data whether or not MMR causes autism? Three, can MMR cause autism?"

This seemingly reasonable suggestion led to a rare disagreement, beginning when Dr. McCormick knocked down the crazy idea that there could be an epidemic of autism. "I don't think the first question is either answerable or in our purview."

"Is there an epidemic of autism?" had already been answered by Reality: A debilitating neurological disorder that struck one in 150 children at the time of their meeting was an epidemic. (Today, autism is now about five times more frequent.)

After acknowledging "one piece of evidence" that supported a vaccine-autism link (the fact that the timing of autism diagnosis often aligned with the introduction of various vaccines), Dr. McCormick stressed again, "We're not saying there is an epidemic." And again, "We are not arguing about whether there is an epidemic."

But they were arguing, and maybe that inspired Dr. Shaywitz to say something really argumentative: "If we were a group working for Philip Morris, we would be saying there's no relation between cancer and smoking. *This* study is not so good, and *this* study is not so good. And the cigarette manufacturers do that constantly... you can pick holes in any study."

He continued: "It's a tough problem, because [autism] is ***hard to define***, and ***nobody has the foggiest idea of what causes it***." They had no difficulty deciding what doesn't cause it, however.

Dr. Bayer called out Dr. McCormick. "Marie, what happened between yesterday and today? I mean it seems to me you have shifted your position." Though shifting one's position is a key attribute in science, Dr. McCormick was quick to deny the terrible notion. "I haven't shifted," and only "would like to be real sure that you all are real comfortable with the quality of the epidemiologic data."

Ronald Bayer

This time, it was Dr. Shaywitz who called out Dr. McCormick: "That's a really leading question, because you never come to vote anyway."

Controversy like this was rare with this group because there were so few occasions anyone said anything challenging or surprising. A standout was the time Dr. Johnston said this:

> "But the question is *What are the data in the <u>other</u> direction*? If you ask for the data that would support a relationship [between vaccines and autism], what are the data that would reject a relationship... ***you are only asking a question in one direction***. That's what the Academy paper is going to do.

> They are going to say, '*Well, because there are no positive data, the answer is that there are not*.'

While he was being so reasonable, it might have been a good time for Dr. Johnston to share some things he'd said at another meeting years earlier:

> "There is very limited pharmacokinetic data concerning ethylmercury. There is very limited data on its blood levels. There is no data on its excretion. It is recognized to both cross placenta and the blood-brain barrier."

> "Aluminum and mercury are often simultaneously administered to infants... there is absolutely no data, including animal data, about the potential for synergy [that would] allow us to draw any conclusions from the simultaneous exposure to these two in vaccines."

But he didn't bother to share that directly relevant and important information with the group.

Dr. Foxman brought up Reye's Syndrome, another blight on Federal public health officials — in that it took FDA many years to warn parents to stop giving aspirin to children, and 25-years before the agency issued an actual ruling.

> Dr. Foxman: In Reye's Syndrome that's not true. They just said, '***Lack of Evidence of an Association***.'

Betsy Foxman

And here Dr. Foxman broke the unspoken rules by being much too clear for comfort, suggesting that they actually answer the question "*Is MMR responsible for an increase in autism cases —* "

An immediate interruption from Dr. McCormick: "That's not the question. The question is, *Is MMR associated*? Increase/decrease has nothing to do with it."

Dr. McCormick's devotion to ensuring nobody affirmed the increase in autism cases is perplexing, given that at the time of her resistance, objections and interruptions, CDC had already determined that the rate was increasing year by year. (CDC's number today is 1 in 36 children, with an undeniable upward trend.)

Surveillance Year	Birth Year	Number of ADDM Sites Reporting	Combined Prevalence per 1,000 Children (Range Across ADDM Sites)	This is about 1 in X children
2020	2012	11	27.6 (23.1-44.9)	1 in 36
2018	2010	11	23.0 (16.5-38.9)	1 in 44
2016	2008	11	18.5 (18.0-19.1)	1 in 54
2014	2006	11	16.8 (13.1-29.3)	1 in 59
2012	2004	11	14.5 (8.2-24.6)	1 in 69
2010	2002	11	14.7 (5.7-21.9)	1 in 68
2008	2000	14	11.3 (4.8-21.2)	1 in 88
2006	1998	11	9.0 (4.2-12.1)	1 in 110
2004	1996	8	8.0 (4.6-9.8)	1 in 125
2002	1994	14	6.6 (3.3-10.6)	1 in 150
2000	1992	6	6.7 (4.5-9.9)	1 in 150

Dodging the (apparently) touchy subject of whether autism had reached epidemic levels, and carefully resolving nothing, the committee members returned to their preferred posture: all for one and one for all.

"A Suggestion That Will Take Us All Off the Hook"

When Dr. Medoff suggested the time-tested IOM strategy of saying nothing and recommending further research, Dr. Kaback didn't agree. "I would worry about that. We do have to make a decision. We don't have the luxury of sitting back and saying, *Yes, well, let's see what happens in a few years.*"

Why exactly was he concerned about waiting? Because "there is a real risk that there will be lots of unvaccinated children. If it turns out that everything has been just a worry that shouldn't be a worry, then there will be two years that go by where you will have lots of unvaccinated people."

Notice that he didn't point out the opposite possibility: If it turned out there *was* reason to worry about vaccine safety, then two years would go by in which millions of children would be injected with unsafe products. (Happens to now be more than 40 years, not just two.)

If a vaccine product was found to be harmful, wouldn't reducing use of that particular vaccine be the desired outcome? Not to the members of this committee, one of whom expressed the foundational concern that if they were to convey too much doubt, "that would have a really major significant impact on people's rates of declines in vaccines."

It's not a crazy idea that a vaccine product might be found undeniably unsafe, as shown by these few examples:

> **Original Salk polio vaccine and Sabin polio vaccine**: Between 1955 and 1963, tens of millions of Americans received polio vaccines contaminated with Simian Virus-40 that contained DNA from African Green monkeys; the SV40 contamination happened twice.
>
> **Sabin Oral Polio Vaccine**: Abandoned in 1999
>
> **Lymerix and ImuLyme (Lyme Disease)**: withdrawn from the market due to serious adverse effects
>
> **OraVax Rotashield (Rotavirus)**: withdrawn from the market for causing intestinal blockage in some infants
>
> **DTP (Diphtheria, Tetanus, and Pertussis)**: some phased out, some withdrawn from market due to adverse reactions including high fever and seizures
>
> **Quadrigen (DPT combined with the Salk polio vaccine)**: withdrawn from market due to safety issues
>
> **Dengvaxia (Dengue Fever)**: caused severe dengue in some people
>
> **Mumps-Measles-Rubella (MMR) Vaccine (Urabe Strain)**: withdrawn from the market due to an increased risk of aseptic meningitis [1]
>
> **Pandemrix (influenza)**: withdrawn because it caused narcolepsy and death in some children

1 

1976 Swine Flu Vaccine: Guillain-Barré Syndrome and other neurological disorders, spinal cord dysfunction, brain dysfunction

Covid Vaccine by Johnson & Johnson: causes blood clotting and death; no longer authorized for use in the US

Covid Vaccine by AstraZeneca: withdrawn from the market, causes fatal blood clots, low platelet counts, thrombosis with thrombocytopenia syndrome

Other vaccines withdrawn or replaced in the market include plasma-derived hepatitis B, HBV Recombinant, whole cell pertussis, killed measles vaccine, tissue-derived rabies vaccine, Lyme, rabies and rotavirus vaccines, and some smallpox vaccines that were withdrawn from the market and then brought back. One smallpox vaccine given to members of the military had the unfortunate result of causing a bunch of people to become infected with a close relative of smallpox called vaccinia. [1]

More recently, it's confirmed that **mRNA Covid vaccines** cause many adverse events including blood clots, cardiac injury and sudden cardiac death in young people. Rather than withdraw these vaccine products, the FDA merely required that a warning be added to future package inserts (that are never seen or read by consumers). Even more recently (and possibly more seriously), both the Pfizer and Moderna vaccine products have been found to contain billions of DNA fragments per dose, unfortunately from Simian Virus 40. [2] [3]

During the writing of this book, FDA has issued warnings about two RSV vaccines, Abrysvo and Arexvy, noting that they cause an increased risk of Guillain-Barré Syndrome (GBS). In healthy people, RSV typically causes mild cold-like symptoms, while Guillain-Barré is a serious neurological disease that causes the immune system to attack the nerves, often affecting the ability to walk and perform everyday tasks independently. Put plainly, we'd all choose to have an RSV infection over Guillain-Barré Syndrome. By the way, rather than withdrawing these products, the FDA merely requires that the new warnings be included in future package inserts (that are never seen by consumers). [4]

---

1  2  3  4 

It doesn't appear that vaccines being unsafe was what mattered most to Dr. Goodman and the group:

> "...clearly we are all sitting around this room, and that represents social expression of concern, and it has come from many sources... what we choose to say and the potential impact, the potential concern, is also of relevance. That is why I was talking about some of the evidence, about the impact of publicity. So we can also have an eye toward what we would hope the impact of our statements would be."

Dr. Gatsonis chimed in with the kind of suggestion the group was sure to love:

> "I have a suggestion that will take us all off the hook."

At first, he thought getting a biologist on the committee would help — but they found better ways to get off the hook: Say what CDC wanted them to say — and avoid saying anything else. With this approach, their conclusions (to use a very generous word) could be promoted in whatever way CDC would later choose.

IOM on Covid Vaccine

In 2023, the IOM convened one of its familiar committees to opine on adverse effects of the new mRNA Covid vaccines. By this point, IOM had rebranded itself with a new and auspicious-sounding name, The National Academy of Medicine. Renamed perhaps, but not recovered from their old wordplay, which the Covid vaccine committee acknowledged right at the start of their report:

> "Readers might be confused by the use of different terms with overlapping meanings, or the same terms to mean different things in different contexts."

To ensure confusion, the 2023 committee decided it would adopt the exact same categories used by the IOM vaccine safety committees in 1991, 1994, and 2012.

Though the 2023 report states that the Committee "was not charged to evaluate the benefits of vaccines," they apparently couldn't help themselves.

> "Covid-19 vaccines are highly effective in adults and children and were key to control of the pandemic.Covid-19 vaccines are estimated to have prevented 14.4 million deaths worldwide in the first year of vaccination alone."

But other than that, they wouldn't evaluate the benefits of the vaccines.[1]

1

CHAPTER SIXTEEN (10 MINS)

"We Are Kind of Caught in a Trap"

"We Have Got a Dragon by the Tail"

Months later, the group convened for the second time, and even then they were still focusing on finding a narrative that wouldn't discourage use of any vaccine no matter what.

Dr. Shaywitz called it like it was:

> "I do think that we have to have ***a consistent story*** if our message is to be heard."

He was convinced that if they didn't mount a clear message, that would disrupt vaccination rates.

> "I mean, we would like to separate the significance of our decision from the science, but people will hear one thing: Is it safe or is it not? And if we waffle, they will say, *well, let's do single dose* [instead of the combination vaccines like MMR] ... and everybody knows that that is going to reduce the vaccinating of children."
>
> Dr. Kaback: I agree with you. I think what is badly needed in this field is some communication mechanisms for health-care professionals who are dealing with vaccinations. Now, the suggestion has been made that ***anything you say about risk is going to decrease adoption of the procedure***. That, of course, is the worry. Then you deal with lowering immunization rates and increasing disease, deaths and problems associated with it ... how do we ensure safety associated with vaccinations? I don't think that notion

> is the right notion. ***We are not trying to ensure safety***. We are trying to maximize safety. We recognize that there may be some risk. This gets back to the probability issue. There are small risks associated with anything. The question is how small are they?"

When applying medical interventions to healthy children, I would have thought the more important question would be, *How big are the risks*? Whichever way the question is asked, these folks were unwilling to answer it.

Dr. Kaback got the group back to what they actually cared about, which was how to say always vaccinate, no matter the product, no matter the child:

> "How do you phrase that? How do you communicate that in the context of a health-care system that gives doctors how long did you say, two minutes or six minutes for the total visit, 1.7 minutes for the vaccination discussion? That is the other big hang up we have here: We are talking about human communication to people with various levels, various ethnic and cultural groups, various abilities to understand statistics probability and scientific information."

Oh, those annoying people of America, with their various ethnic groups and various abilities to understand scientific information.

> "We have got a dragon by the tail here. At the end of the line, what we know is —and I agree— that the more negative that presentation is, the less likely people are to use vaccination."

> "We are kind of caught in a trap. How we work our way out of the trap, I think, is the charge."

Actually, the charge was to evaluate the evidence for "possible causal associations between immunizations and certain adverse outcomes," but assignments change, apparently, when they become a trap.

Dr. Bayer opened Pandora's Box of Semantics to see if maybe they could redefine the word that gave them the most trouble. "The word *safe* is interesting," he said. "Safe is not a scientific term. Safe is a social judgment. It always means safe as compared to what? When you build a bridge, you build in a safety factor, knowing that ***you could always make it safer if you spent more or did more***."

Dr. Bayer's observation that "the word safe is interesting" is itself interesting given the word's role in the most popular phrase of the medical establishment: *Safe and Effective*. Despite Dr. Bayer's complicated idea that safe is a social judgment, most Americans probably see the issue more simply. Asking if a product is safe to inject into one's child means safe in the regular sense of the word, as in *free from danger or injury*; not *exposed to danger or harm.*

Reading the gyrations of the IOM committee, I'm reminded of a routine by the great comedian George Carlin. Ridiculing deception in advertising, he recited the phrase *real chocolatey goodness.*

"You know what chocolatey means? No f-ing chocolate!"

The job of IOM's Immunization Safety Review Committee was "to evaluate the evidence of possible causal associations between immunizations and certain adverse outcomes."

You know what 'evaluate the evidence' means? No f-ing evaluation.

"The committee begins from a position of neutrality."

You know what 'position of neutrality' means? No f-ing neutrality.

> Dr. Kaback: I have heard multiple times already about the issue of autism with MMR. It is building... we live in a 24-hour-a-day, seven-day-a-week media blitz on our public. They are always looking for things to fill up the time and the newspapers with. This makes a great story. So, we have a big job.

The big job he identified was that of countering the growing concern that autism might be linked to the MMR vaccine. Again and again, the lauded scientists made clear that their charge was not to study that issue, but to kill that issue.

> Dr. McCormick: I took away actually an issue that we may have to confront, and that is actually the definition of what we mean by safety. It is safety on a population basis, but it is also safety for the individual child. I am wondering, if we take this dual perspective, we may address more of the parent concerns, perhaps developing a better message if we think about what comes down the stream as opposed to **CDC, which wants us to declare, well, *these things are pretty safe*** on a population basis.

Dr. Goodman was hesitant to give parents any insight into possible risk:

> "I think we have to be very, very careful... quantifying the impact of the side effects, I just think that we are going to have to be very careful. The very use of it will be seen by some, partially legitimately and I think partially illegitimately, as an ideologic bias of the committee."

He was right that some people would perceive an ideological bias in the committee. I am one of those people. The IOM committee would ultimately conclude that every vaccine should be given to every child no matter what. The really important issue, according to Dr. Johnston, was that "the usage of MMR is dropping away." And that's what they had to fix.

Though Dr. Shaywitz had said they would like to "separate the significance of our decision from the science," they certainly didn't tell anyone that's what they did. In the end, they presented their decision, to the extent there even was a decision, as if it was science — when actually, it was political science.

> **political science** *(noun)* — the description and analysis of political and governmental institutions and processes

Agent Orange, Again

The closed-door sessions had certainly closed the door on every possible outcome other than the predetermined one: Disregard or minimize any evidence that any vaccine could possibly cause or contribute to autism in any child, and debunk that claim — not by science but by wordplay.

And what better wordplay than that used in past Government debunkments? Dr. Goodman, the master of categories:

> "I do think the Agent Orange categories work well, and we can decide how we work them... have a category that simply says *Causality Suggested* but it is *Not Sufficient*."

To Dr. Goodman's credit —and it hasn't been easy to extend much credit to him— he acknowledged that observational evidence can indeed be used to establish causation. "There is observational evidence like tobacco and lung cancer that we do, over time, actually accept as sufficient for establishing a causal relationship, combined with the biology."

Having invoked the deadly association between tobacco and lung cancer, Dr. Goodman revisited Agent Orange, paying homage to how the Government craftily crafted the crafty language that would best say nothing: "The Agent Orange standards were *Insufficient Evidence,* that is, inadequate to make a claim either way, and *Limited/Suggestive Evidence.*"

Can anyone tell the difference between categories used by IOM when opining on the bioweapon versus the categories they used for childhood vaccines?

> **"Limited or Suggestive Evidence of an Association"**
> **"Inadequate or Insufficient Evidence to Determine an Association"**
> **"Not Showing a Positive Association"**
> **"Inadequate/Insufficient Evidence to Determine Whether an Association Exists"**
> **"Limited or Suggestive Evidence of No Association"**

Are these phrases from IOM committees on vaccine safety, or Agent Orange?

Answer: Both.

"Definitive Proof That It Doesn't Not Cause Autism"

Guided away from the issue of an increase in autism, Dr. Johnston came back to an alternative inquiry: "The question is, *Can MMR cause autism*?"

After that momentary flash of authentic exploration, the group slipped back to their default: Composing language, naming categories, and opining on how they should say things.

> Dr. Gatsonis: If you said the following statement, that *the presently available information* —and by presently available, we mean information that has been understood, preferably— and so on and so forth, *does not indicate that there is a causal relationship*. The lead statement would be *the presently available and verified information indicates that*... If you use some words like this, at least you are saying that based on everything that we heard yesterday that we think we can trust —because there were things that were said that we haven't read— we haven't seen, we cannot trust...

And around the parsing track they went again, with someone speculating they were now at a Level "Two Minus."

> Dr. Gatsonis: No, that makes it a —
>
> Dr. Shaywitz couldn't help but interrupt: "It favors the rejection of a causal relationship."

Just as the group might have been circling the decision drain, Dr. Medoff resurrected his favorite issue: "The problem is the ***wording*** of the categories," to which Dr. Johnston asked, perhaps sarcastically: "Maybe you want to change them?"

> Dr. Medoff: We are put in a position of not being able to reject the hypothesis of whether or not MMR causes autism, because there is no definitive proof that it doesn't cause autism. Basically, isn't that what you are saying, that there has to be definitive proof that ***it doesn't not cause autism***?

Triple negatives aside, Reality cleared things up:

> Dr. Johnston: ***There doesn't have to be proof, because you will never really have it.***

Acknowledging there would never be proof of their position, Dr. Stratton said all they needed was "one or more epidemiologic studies that say there is not a relationship."

> Dr. Shaywitz: It seems to me we can say at this point —I like Constantine's phrasing— There is *No indication at this time,* and so we *Favor rejection,* however the wording is.

Dr. Shaywitz said their problem was "the same as everybody has faced writing all of these [IOM reports] … there could be a negative study somewhere that is in somebody's file drawer that never got to publication, that was ***blocked by JAMA, not to publish***."

> Dr. Parkin: I'm just concerned that we be careful about how we talk about the quality of the studies … because if we say that there is no association, and we haven't addressed the issue as to whether the studies were ***sufficiently designed and conducted to find an association if one exists,*** then we are going to have problems after the report is released.

Hi Reddit! I'm Rebecca Katz, PhD MPH & Professor at Georgetown University.
I've spent 20 years working on pandemic preparedness.
AMA about COVID-19

Rebecca Parkin

And right she was.

> Dr. Bayer: As I read those studies, they simply answer the question of whether there has been, on a population basis, an increase in autism that is connected to the introduction of MMR. They don't ask the question, *Does MMR cause autism*? If they can't, then you can't think of evidence to reject them.

And right he was.

Vernice Davis-Anthony

> Vernice Davis-Anthony: While we are firm on our conclusions based on the evidence, there are other issues that others may consider or that need to be considered. Make it real clear that we are not wavering on our

> conclusions based on the evidence, because we're not. We're real clear on that. And ***we do not want to give the impression that we do not understand the complexities of autism, that we don't have all these answers.***

And right she was, I think.

> Dr. Berg: One of the criticisms of the IOM products is that there is a lack of consistency from panel to panel about how they approach issues and about the explicitness and transparency of their methods.

And right he was.

Dr. Berg expressed another concern through an anecdote:

> "I was at a meeting in San Diego about a month ago about prostate cancer. The conclusion of the scientists is, *we don't know what causes it, we don't know how to prevent it, we don't know whether screening works, we don't know if treatment works*. The CDC says, *Yes, but that is not helpful. What should we actually tell people to do*? The scientists say, *the answer is we don't know*. The CDC says, *yes, I know, but now what do you really think?*"
>
> "I personally end up feeling like it is okay to conclude that we don't know, and end there."
>
> "CDC is, in my view, one of the malefactors on this issue, because they are always pushing to go beyond the data and say *Yes, but*."

And right he was.

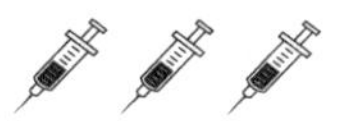

In addition to the contortions of their semantic circus acts and their pretense of science, the members of IOM's committee used another strategy: rushing their report to ensure it came out before a number of actual biological studies they knew were imminent. They also declined to review those studies in advance, because what you don't see is often quite useful, like when you don't see evidence of risk or harm.

The biological studies they were able to miss came from scientists at Columbia University, University of Arkansas, Northeastern University, Johns Hopkins University, Harvard University, and the University of Washington. One demonstrated that injected thimerosal ended up accumulating in the brain, and made reference to the IOM's committee: "This approach is difficult to understand, given our current limited knowledge of the toxico-kinetics and developmental neurotoxicity of thimerosal, a compound that has been and will continue to be injected in millions of newborns and infants." [1]

Speaking of injecting thimerosal into babies, the IOM report is often cited as proof that thimerosal is "safe," but the report never makes that claim. In fact, IOM Committee Chair Marie McCormick has made an opposite admission:

> "The committee accepts that under certain conditions, infections and heavy metals, including thimerosal, can injure the nervous system." [2]

Something that injures the nervous system is not safe. That fact is clear from thimerosal's ironically named Safety Data Sheet:

> Not for medicinal use
> May cause damage to organs
> Toxic if swallowed
> Fatal in contact with skin
> Fatal if inhaled [3]

IOM didn't worry about those warnings, because after all, it was only mercury being injected into children, not Agent Orange, for God's sake. And even if it had been Agent Orange, IOM had already concluded that's not so bad anyway.

---

1 2  3 

CHAPTER SEVENTEEN (6 MINS)

Trust the Media?

A company called Theranos developed (and intensely promoted) a revolutionary blood-testing technology that, according to Theranos, could perform hundreds of medical tests with a single drop of blood. The charismatic founder and CEO of the company was Elizabeth Holmes.

Contrary to the company's lofty business projections, a jury convicted Elizabeth Holmes of medical fraud in 2021. During the trial, she admitted that her company conducted medical tests using the same old machines that were supposedly being replaced by her remarkable new technology.

Though people were given fraudulent and incorrect medical results, the news media's fawning over Holmes and Theranos kept the fraud going strong for years, such that the company was eventually valued at $9 billion. Not a typo.

Theranos would not have been possible if the news media were skeptical or even mildly curious when it comes to new medical technology and pharma products. As you scan these few examples of news media promotion, consider that Vioxx and opioids and Covid vaccines all floated into our society on a similar cushion of media praise, free of scrutiny or skepticism.

USA Today:

> *Elizabeth Holmes is tall, smart and single. Well, maybe not truly single. "I guess you should say I'm married to Theranos," Holmes says with a laugh. Only she's not kidding... while Holmes is a billionaire on paper, nothing seems to interest her less... "We're successful if person by person we help make a difference in their lives," says Holmes, who has a soft yet commanding voice that makes a listener lean in as if waiting for marching orders.*

New Yorker:

> *Although she can quote Jane Austen by heart, she no longer devotes time to novels or friends, doesn't date, doesn't own a television, and hasn't taken a vacation in ten years... "I have done something, and we have done something, that has*

changed people's lives... I would much rather live a life of purpose than one in which I might have other things but not that."

CNN:

The company she founded has the potential to change health-care for millions of Americans.

FORBES:

Elizabeth Holmes, 30, is the youngest woman to become a self-made billionaire — and she's done so four times over. "What we're about is the belief that access to affordable and real-time health information is a basic human right, and it's a civil right," she says.

Since the news media couldn't change course after all that worship, it wasn't a journalist who undid the Theranos scam. It was a Stanford professor named John Ioannidis who was willing to publicly point out that Theranos hadn't published any peer-reviewed research about their products.

(To bring us to the present moment for a moment, John Ioannidis is an esteemed physician, scientist and epidemiologist who was among the first and most vocal critics of lockdown policies. For that, he's been the target of excoriation and cancellation by the news media, which silently screamed, *Don't you dare wake us from our Covid fever dream, Professor!*)

While still being loved up by the news media, Theranos boosted its credibility by getting then-Vice President Biden to visit their facility. In order to conceal the lab's true operating conditions, Holmes and her team created a fake lab for the Vice President to tour.

The Theranos case should remind us that 25% of drugs approved by the FDA are later pulled from the market — so they can't all be miracle drugs. You wouldn't know that, however, by the breathless news media excitement that greets each new product. Whatever a pharma company or public health bureaucrat says, the news media repeats, enshrines as fact, and then defends. Never has the media's lack of skepticism and curiosity been on such colorful display as it was in recent years, during which they constantly promoted and parroted false pharma claims related to new vaccines (100% efficacy, safe and effective). Same for false claims about Pfizer's Covid treatment, false claims about Ivermectin, and every other false claim by Pharma and government. Real journalism being MIA left Americans living in a world of compliance rather than science.

Can news media companies be trusted? Oh yes, they can be trusted to serve their Pharma masters. The pharmaceutical industry is the biggest sponsor for television news programs, accounting for billions of dollars in revenue every year. Pfizer alone spends nearly $3-billion a year on advertising. The majority of pharmaceutical products advertised on television cannot be purchased or selected by viewers without a prescription, meaning the massive advertising spend must have benefits other than influencing consumer behavior. Indeed, the billions spent by Big Pharma literally support television news media departments — and guarantee the abject avoidance of any critical news stories about vaccine products. Not once in recent years did any television news interviewer pose a single difficult question to Pfizer CEO Albert Bourla, or Anthony Fauci, or President Biden. Hundreds of interviews, and not one difficult question challenging the new vaccine products.

The US Government project to ensure global mass-vaccination with new mRNA products was among the largest endeavors in human history, with more than $10-billion spent on persuasion campaigns alone. As such, one might expect attentive news media scrutiny and analysis of everything to do with the new products. That didn't happen. Not even close.

In June of 2021, a massive Government-funded media campaign pushed the message that the new mRNA vaccines were safe and effective. Americans were being hammered with the message that mRNA vaccination was the only way to resume a normal life, the only way to dine in a restaurant, fly on an airline, study at a university, be welcomed at a hospital, attend a show or concert, or feel good about yourself by meeting what was promoted as a civic duty.

On June 16th of that year, a previously healthy 15-year-old Connecticut boy was found dead in his bed two days after receiving the second dose of the Pfizer Covid vaccine.

On the very same day, a previously healthy 13-year-old Michigan boy was found dead in his bed three days after receiving the second dose of the Pfizer Covid vaccine.

An autopsy in Connecticut and another in Michigan confirmed the official cause of death for both boys: ***Vaccine-induced myocarditis*** (swelling of the heart caused by the vaccines). [1]

---

1 

When government medical examiners in two states determine that Pfizer mRNA vaccines killed two healthy teenagers, smack in the middle of a mass vaccination program with Pfizer's product, wouldn't that seem to be an important and obvious story for national news media? Apparently not in the opinion of the national news media — for none of the cable news channels reported the deaths from the new Pfizer product — and neither did CBS, NBC, or ABC News.

Similarly, when sudden deaths occurred in front of spectators at high school sports events, meaning a teenager collapsed dead on the basketball court, for example, no reports in national media.

Previous to 2021, such events involving a sudden death of a young athlete would keenly focus on answering the glaring question: *Why did this person die?* But a new reporting protocol emerged during Covid: Ignore such stories outright. In the vanishingly rare instances that these deaths were even mentioned beyond local news outlets, the stories contained phrases most readers have now subconsciously come to expect and accept:

> *Cause of death unknown, died of natural causes, no cause of death has been given, died after a brief illness, officials are still investigating cause of death.*

No follow-up news stories, no questions from reporters, no curiosity.

Healthy young people dying suddenly in the middle of play, or during sporting events, or while sitting at the dining table with their families, or while sleeping — these used to be very rare events. Most of us never heard of such events during our youth or our entire lives. But in recent years, sudden deaths among healthy young people have occurred frequently. (See Appendix #2 for a mere sampling of more than 250 such cases in just 2021 and 2022.) The fact that so many unexpected deaths occurred among healthy young people should have been explored in a thousand news reports — and wasn't, which leads to a quick thought experiment:

> *Imagine that thousands of healthy young Americans died suddenly, unexpectedly, mysteriously — and then kept dying at an alarming and escalating rate. (Once upon a time), that would trigger an urgent CDC inquiry to determine the cause of the deaths. Imagine attentive and curious public health officials discover the decedents had all repeatedly ingested a new and little-understood drug. Next, the officials determine to a certainty that the drug these kids took has a clear mechanism of action that can cause inflammation of the heart and other cardiac injuries. They learn that public health officials in other countries have seen the same thing and stopped recommending that drug to young people. Next, some of the most senior and revered scientific advisors to the US Government*

> *publicly recommend the drug be stopped for young people. Finally, thousands of doctors around the world sign petitions and write op-eds opposing the drug for young people. Experts from Harvard, Yale, MIT, Stanford, and Oxford universities come forward to voice their concerns.*

Alas, that thought experiment doesn't require any imagination, because it's exactly what occurred — except for the part about attentive and curious CDC officials rushing in to inquire. That part I had to make up.

In the pre-Covid world, wouldn't inquisitive reporters chase such a story, and wouldn't the FDA pause administration of the new mystery drug until a comprehensive inquiry was complete?

And above all, ***wouldn't such a drug have quickly become a leading suspect worth considering for its possible role in the thousands of deaths***?

Somehow, those have become rhetorical questions.

Example: The *Wall Street Journal* ran a story about a lethal conundrum facing insurance companies. Apparently, as mass vaccination progressed in 2021, excess death claims in working-age Americans *tripled*. Given the temporal relationship, might there be a connection between mass vaccination and these deaths? Apparently not, because when *WSJ* dug into what might have caused these deaths, mass vaccination wasn't even among the possibilities worth considering:

1. Delayed medical treatment from 2020
2. People's fear of seeking treatment
3. Trouble lining up appointments
4. Drug abuse and other societal troubles
5. People not taking care of themselves
6. Long Covid
7. Not-yet-known long-term effects of Covid
8. People dying later "from the toll Covid has taken on their bodies"

Numbers 1, 2, and 3 are all subsets of the same concept: impact of lockdowns and fear. Numbers 6, 7, and 8 are all subsets of the same concept: the impact of Covid illness.

Aside from drug abuse and trouble getting doctor appointments, did anything else happen in 2021 that might-possibly-maybe-perhaps-call-me-crazy be worth considering?

The insurers knew that most of the 2021 excess deaths among their customers were due to heart and circulatory issues, neurological disorders, and stroke. Coincidence, I guess, that nearly all their deceased customers had just been injected with a new drug known to cause heart and circulatory issues, neurological disorders, and stroke.

The *Wall Street Journal* and other news media outlets faced a real head-scratcher: Could any of those unexpected deaths possibly be linked to that new, never-before-used, minimally tested, maximally rushed, mass-injected drug ***known to cause the very medical issues these working-age Americans were dying of***?

But the *Wall Street Journal* didn't ask this obvious question. No major news media company asked any important questions at any time during the entire mass-vaccination campaign. Are they all gullible in their acceptance of anything offered up by Big Pharma and Government? Were they fooled and hoodwinked by Big Pharma and Government?

No, they were complicit.

WSJ

MARKETS

Rise in Non-Covid-19 Deaths Hits Life Insurers

By Leslie Scism Follow
Updated Feb. 23, 2022 5:36 am ET

THE HILL

Well-Being > Longevity

'Huge, huge numbers:' insurance group sees death rates up 40 percent over pre-pandemic levels

By Shirin Ali | Jan. 7, 2022 | Jan. 07, 2022

'Just Unheard Of': Another Insurance CEO Admits Unexplained Deaths Are Up 40% Among Working People

Fact checked

July 4, 2022 Baxter Dmitry News, US 11 Comments

Why Are All-Cause Excess Deaths in the Under-45s So Much Higher This Year Than at the Height of the Pandemic?

BY NICK RENDELL 15 AUGUST 2022 2:00 PM

SHARE

news.com.au

Excess deaths in 2022 'incredibly high' in Australia

The Australian government should be urgently investigating the "incredibly high" 13 per cent excess death rate in 2022, the country's peak actuarial body says.

Frank Chung
@franks_chung 6 min read December 8, 2022 - 9:42AM news.com.au

Most corporate news media companies decided to *not* publish stories on excess deaths among working-age Americans. For a look at other important information the news media intentionally underreported, see Appendix #4 called "NEWS YOU LIKELY MISSED," on Page 186. Just one example for now: News media companies did not bother to be sure Americans knew that mRNA vaccines cause cardiac problems. (If you have any doubt about this, you can see a sampling of published scientific papers in Appendix #1 on Page 161.)

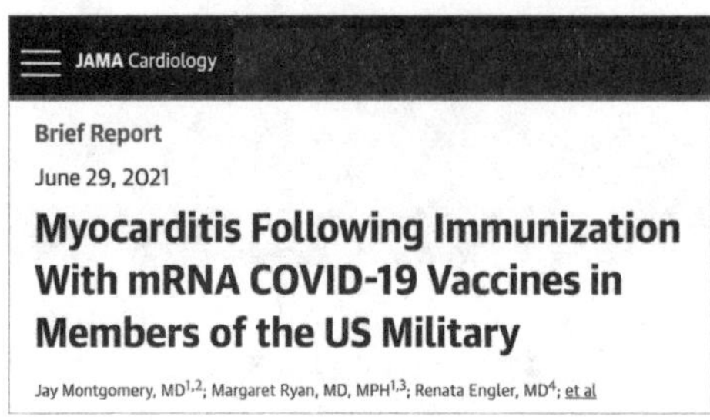

JAMA Cardiology

Brief Report

June 29, 2021

Myocarditis Following Immunization With mRNA COVID-19 Vaccines in Members of the US Military

Jay Montgomery, MD[1,2]; Margaret Ryan, MD, MPH[1,3]; Renata Engler, MD[4]; et al

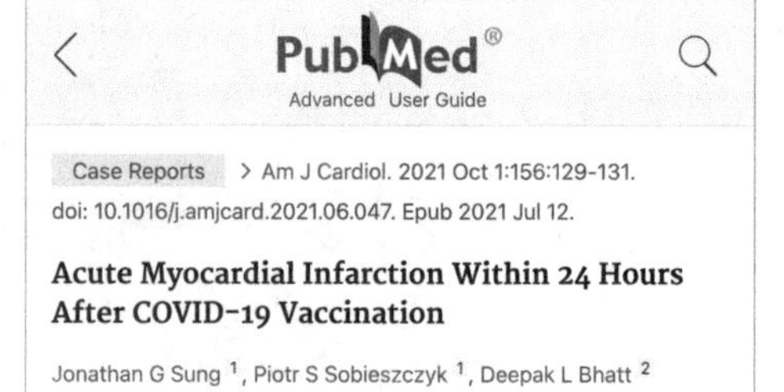

PubMed®

Advanced User Guide

Case Reports > Am J Cardiol. 2021 Oct 1:156:129-131. doi: 10.1016/j.amjcard.2021.06.047. Epub 2021 Jul 12.

Acute Myocardial Infarction Within 24 Hours After COVID-19 Vaccination

Jonathan G Sung [1], Piotr S Sobieszczyk [1], Deepak L Bhatt [2]

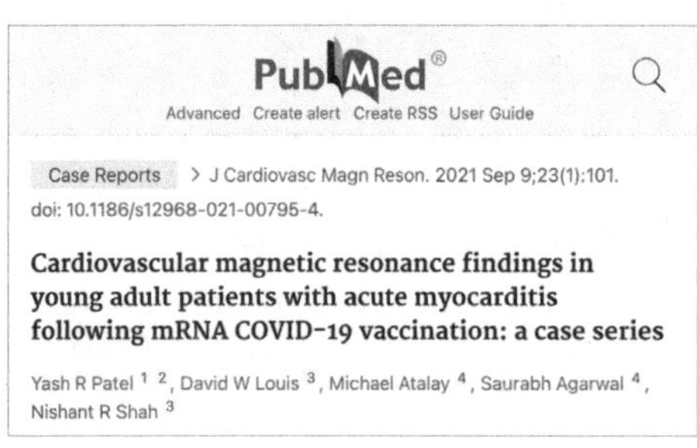

PubMed®

Advanced Create alert Create RSS User Guide

Case Reports > J Cardiovasc Magn Reson. 2021 Sep 9;23(1):101. doi: 10.1186/s12968-021-00795-4.

Cardiovascular magnetic resonance findings in young adult patients with acute myocarditis following mRNA COVID-19 vaccination: a case series

Yash R Patel [1 2], David W Louis [3], Michael Atalay [4], Saurabh Agarwal [4], Nishant R Shah [3]

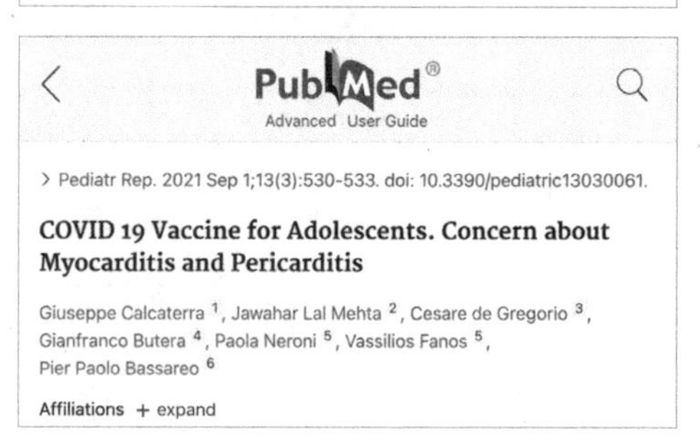

PubMed®

Advanced User Guide

> Pediatr Rep. 2021 Sep 1;13(3):530-533. doi: 10.3390/pediatric13030061.

COVID 19 Vaccine for Adolescents. Concern about Myocarditis and Pericarditis

Giuseppe Calcaterra [1], Jawahar Lal Mehta [2], Cesare de Gregorio [3], Gianfranco Butera [4], Paola Neroni [5], Vassilios Fanos [5], Pier Paolo Bassareo [6]

Affiliations + expand

eClinicalMedicine
Part of THE LANCET Discovery Science

Cardiac manifestations and outcomes of COVID-19 vaccine-associated myocarditis in the young in the USA: longitudinal results from the Myocarditis After COVID Vaccination (MACiV) multicenter study

Supriya S. Jain [a] · Steven A. Anderson [b] · Jeremy M. Steele [c] · Hunter C. Wilson [d] · Juan Carlos Muniz [e] · Jonathan H. Soslow [f] · et al. Show more

Newsweek

More Study Urged After 7 Teen Boys Suffer Heart Inflammation After Getting COVID-19 Vaccine

Published Jun 04, 2021 at 1:20 PM EDT

Open Access Article

Myocarditis and Pericarditis following COVID-19 Vaccination: Inequalities in Age and Vaccine Types

by Minghui Li [1,†], Jing Yuan [2,†], Gang Lv [3], Jacob Brown [4], Xiangxiang Jiang [5] and Zhiqiang Kevin Lu [4,*]

European Heart Journal
ESC European Society of Cardiology

JOURNAL ARTICLE EDITOR'S CHOICE

Booster vaccination with SARS-CoV-2 mRNA vaccines and myocarditis in adolescents and young adults: a Nordic cohort study

Anders Hviid, Tuomo A Nieminen, Nicklas Pihlström, Nina Gunnes, Jesper Dahl, Øystein Karlstad, Hanne Løvdal Gulseth, Anders Sundström, Anders Husby, Jørgen Vinsløv Hansen ... Show more

European Heart Journal, Volume 45, Issue 15, 14 April 2024, Pages 1327–1335,

Newsweek

News | Health And Medicine Florida Youth Health Care Vaccine Doctor

Top Florida Doctor Warns Young Men COVID Vaccines Pose 'High Risk' of Death

Published Oct 08, 2022 at 5:46 PM EDT

CHAPTER EIGHTEEN (1 MIN)

If I Gave My Child All the Recommended Vaccines, Was That a Mistake?

No, it was not a mistake. Based upon the information put in front of you at that time, it was probably the obvious choice and the best decision. The word *decision* shares a root with *incision,* and means to cut off from all other possibilities. Most of us were never even aware there were other possibilities. There was no counter-argument about vaccines, because there was no argument in the first place. Vaccination was simply one of the obvious steps, starting on the day of birth and continuing from then onward.

We feed, we comfort, we protect, ***we vaccinate***. That is early parenting as most people know it — nearly automatic.

Our decisions going forward, however, can be improved by broadened awareness; we now know there is a lot to consider when it comes to selecting which vaccines we'll give and which we'll avoid. Experts have studied the topic, and parents can do the same. Just as different experts have different opinions, so are parents entitled to have different opinions. Experts are rewarded for embracing the orthodox view, and punished for rejecting it. But parents are able to be more objective in their decision-making.

Experts are taught, accepted into an industry, and granted a license; that's how they gain authority. ***Parents have inherent authority from Nature.***

Decisions made in the past are beyond our reach. The only decisions that matter are the ones we make from now forward. You now have a great deal of information to help decide which vaccines make sense to you, and which, if any, don't make sense — to you.

CHAPTER NINETEEN (2 MINS)

Ask Your Doctor

Some people invest confidence into doctors because of their medical degree and medical license; others invest confidence into specific doctors based upon how the doctor responds to important questions during a personal meeting. An in-person discussion is key because if asking questions via email or text, the doctor's answers might have been acquired via a web-search, or maybe just a visit to the CDC website — and that we can do on our own.

It's fair to expect a pediatrician to know very important facts without checking the CDC website. For example:

Can any vaccines cause autism in any children?

If the doctor assures you that the notion has been debunked, you might ask how it was debunked and by whom? (You know the answer from Chapters One and Two.)

What do you believe causes autism?

Why do you think that?

Is it true that thimerosal (ethylmercury) has been removed from all childhood vaccines?

Are there any vaccines still given to children that contain thimerosal?

Which vaccine products (by brand name) contain thimerosal, in case I want to avoid those?

Have you read the package inserts for the vaccines you administer?

Do these package inserts acknowledge any neurological injuries as possible adverse events from their vaccines?

I've always wanted to read the package inserts; may I see them or have copies?

Which of these ingredients are contained in vaccines given to children? (Put an X through any that are *not* vaccine ingredients.)

Gelatin from boiled pig skin Chicken embryo protein

Blood from the hearts of cow fetuses Human fetus DNA fragments
Albumin from human blood plasma
Oil extracted from shark livers
Insect cell proteins
Monkey kidney DNA particles
Formaldehyde
Polysorbate 80
Potassium chloride
Phenol
Borax/sodium borate
Monosodium glutamate (MSG)
Aluminum salts Thimerosal/ethylmercury
Triton X-100

If a child is injured by a vaccine, can the family sue the vaccine manufacturer?

If a child is injured by a vaccine, what government agency compensates the family?

Has the US Government ever compensated any family for autism caused by a vaccine?

Has the US Government ever compensated any family for any neurological injuries or brain damage caused by a vaccine?

Have you personally ever seen any adverse reactions to any childhood vaccines?

Can the BCG vaccine be beneficial to health beyond protection from tuberculosis?

What is the main adverse effect associated with human exposure to mercury?

See Appendix #3 on Page 182 for a cheat-sheet of accurate answers.

CHAPTER TWENTY (3 MINS)

Crimes & Criminals: RICO

So, you now know how the vaccine-autism link was debunked by the Institute of Medicine at the ~~request~~ behest of CDC, which was acting at the behest of the military and Big Pharma, which were acting at the behest of their biggest customer (the US Government), which was acting at the behest of the lobbyists and politicians engaged by Big Pharma, which was acting at the behest of...

As that snake consumes its tail, Americans continue to be told and sold what to think. Since 2021, Americans have been persuaded to take the ~~recommended~~ required Covid injections: one to start, another a few weeks later, followed by another some months after that, followed by more in Year One, followed by at least one each year thereafter, for life. Eventually, Pharma succeeded at getting three mRNA injections to be required for all babies starting at six months old. They never even had to make a case as to why that might make sense.

When a consumer is injured by a product, the obvious answer would be a lawsuit against the manufacturer. I say *would be* because the National Childhood Vaccine Injury Act of 1986 and the PREP Act of 2005 protect vaccine-makers from any liability for injuries their products might cause. No matter how reckless the manufacturer, no matter how toxic the injected material, no matter how grievous the injury, these lucky companies are inoculated against lawsuits. [1]

The vaccine-makers cruise the Mediterranean in their yachts, the public health officials retire and take jobs with the vaccine-makers, the media companies make a fortune from Pharma, their biggest sponsor, while the families that saw their children regress after vaccination, and those people injured by the Covid mRNA vaccines, for example — those people have no recourse.

1

As a criminologist, it would be hard for me to miss the interlinked companies and individuals and institutions and agencies that are working together to create and protect their gruesome gravy train. It looks like organized crime — because it is organized crime.

The Racketeer Influenced and Corrupt Organizations Act of 1970, known as RICO, is a Federal law that provides extended criminal penalties for acts performed as part of an ongoing criminal enterprise.

RICO was originally used to prosecute the Mafia and other groups engaged in organized crime, meaning literally, crime that is organized by conspirators. A recent example: Some Pharma executives, pharmacies and doctors concocted a scheme getting doctors to write prescriptions for profitable products, and then referring their patients to co-conspiring pharmacies. Then the hack healers split the profits. [1]

> *"We knew that if we got a patient on the drug, over time they would become a greater revenue stream. The longer the patient stayed on the drug, the higher the dose that they were going to use, the more revenue it was going to be worth to us."*
>
> — Michael Babich, CEO
> Insys Therapeutics

Another RICO case involved Insys, a company that conspired to bribe doctors to prescribe its fentanyl-based pain medication to patients who didn't need it, some of whom died. The more common Pharma conspiracy is to pay doctors to "educate" other doctors about specific products, but an ambitious Insys employee proposed they could just pay doctors directly to write masses of unneeded prescriptions. *Take the bribe and prescribe.* When the company's founder heard about his employee's scheme, he slammed his fist on his desk and said, "That's our next VP of sales!" And right he was.

> **conspiracy** *(noun)* — when two or more people secretly agree to commit a harmful act

1 

Because the phrase *conspiracy theory* is so often weaponized to make such things sound crazy, you'd think conspiracies are rare. They are not rare. Conspiracies are very common.

For example, imagine a company is planning to launch a new product. Executives decide to overstate its effectiveness. At the same time, they make plans to disparage competitive products in unfair and untrue ways. And naturally, they keep all this secret. That is a conspiracy.

Might various stakeholders in Pharma, regulatory agencies, medical companies, politics and media have conspired together to use fraud, artifice, data manipulation and data destruction to produce misleading studies that exaggerate the efficacy and safety of some products? Might aligned parties have used medical associations, universities, industry front groups, and an army of trolls and bots and public officials to harass, punish, deceive, intimidate, discredit, defraud, silence and terrorize any doctor, parent, scientist, elected official, or podcaster who dared to question vaccine orthodoxies (as Moderna did to podcaster Russell Brand and other critics [1])?

Might they have systematically gained control over scientific journals, mainstream media, medical schools, hospitals, and trade organizations, and then used these assets to deploy propaganda and deceitful claims in order to increase their profit and power?

The answer to all those questions is Yes — and that describes a RICO case.

Reading this book, ***you are the jury*** — able to clearly see a corrupt and powerful and well-funded enterprise that's been stridently protecting the highest-grossing products in world history by concealing and distorting anything unfavorable about their drugs, and marginalizing or destroying anyone who raises real questions.

As a member of this unique jury, you now have more insight on these topics than just about any supposed expert you'll ever encounter. As such, you can invite others to join you in this awareness, to seek their own best understanding, and to acknowledge that what we inject into children is not so simple as the people in charge have wanted everyone to believe.

1

CHAPTER TWENTY-ONE (1 MIN)

Who to Trust

You've now read a lot about vaccines, autism, childhood diseases, public health officials, federal agencies, Big Pharma, bribes, payoffs, regulatory capture, the news media, criminal fines, side-effects, vaccine injuries, deaths, children, and IOM committees.

You've seen up close the process that ostensibly debunked the vaccine-autism link. It's your call if you can trust that process or the people hired to act it out.

As detailed in the book *VACCINE WHISTLEBLOWER, Exposing Autism Research Fraud at the CDC*, public health officials carefully hid the vaccine-autism link that their own research had confirmed. Dr. William Thompson, the senior CDC official who admitted the coverup and became a whistleblower:

> *I regret that my coauthors and I omitted statistically significant information in our 2004 article published in the journal* Pediatrics. *The omitted data suggested that African American males who received the MMR vaccine before age 36 months were at increased risk for autism. I'm completely ashamed of what I did… the higher-ups wanted me to do certain things, and I went along with it.*[1]

At the height of the coverup, the director of CDC was Dr. Julie Gerberding. After the good doctor left CDC, she became president of Merck's multi-billion-dollar vaccine division. (Merck is the company that makes the MMR vaccine, if you haven't already guessed that.) It's your call if you can trust CDC. Or FDA. Or Merck.

Throughout these pages, you've had the opportunity to check citations and consider original source material, including a lot of reassuring information about vaccines, advanced by Federal public health officials and Pharma companies.

Who to trust?

---

1 

Trust yourself. Trust yourself because even though these twenty-one chapters have been fairly brief, you've almost certainly read more about how vaccine harms were debunked than any doctor you'll ever meet, including pediatricians.

Trust yourself.

Appendix #1

Sampling of Published Papers on Covid Vaccine-Induced Cardiac Injuries to Young People

1. Covid-19 Vaccine for Adolescents. Concern about Myocarditis and Pericarditis - PubMed https://pubmed.ncbi.nlm.nih.gov/34564344/
2. Be alert to the risk of **adverse cardiovascular events after Covid-19 vaccination** https://www.xiahepublishing.com/m/2472-0712/ERHM-2021-00033
3. **The vaccine was associated with an excess risk of myocarditis** https://www.nejm.org/doi/full/10.1056/NEJMoa2110475
4. **Cardiac Injury in Adolescents Receiving the Covid-19 Vaccine** https://www.ncbi.nlm.nih.gov/pubmed/34077949
5. **Myopericarditis After the Pfizer Messenger Vaccine in Adolescents** https://www.ncbi.nlm.nih.gov/pubmed/34228985
6. Myocarditis and pericarditis after Covid-19 vaccination: "**The incidence rate was higher in adolescents and after the administration of the second dose of messenger RNA (mRNA) vaccines. Overall, mRNA vaccines were significantly associated with increased risks for myocarditis/pericarditis.**" https://www.mdpi.com/2075-4426/11/11/1106
7. Clinical suspicion of **myocarditis temporally related to Covid-19 vaccination** in adolescents and young adults https://www.ahajournals.org/doi/abs/10.1161/CIRCULATIONAHA.121.056583?ub%20%200pubmed
8. The relatively low risk posed by acute Covid-19 in children, and uncertainty about the relative harms from vaccination and disease mean that the balance of risk and benefit of vaccination in this age group is more complex. (Andrew Pollard is chair of UK's Joint Committee on Vaccination & Immunization) https://pubmed.ncbi.nlm.nih.gov/34732388/
9. Epidemiology and clinical features of myocarditis/pericarditis before the introduction of Covid-19 mRNA vaccine in Korean children: a multicenter study https://search.bvsalud.org/global-literature-on-novel-coronavirus-2019-ncov/resource/en/covidwho-1360706
10. Myocarditis associated with SARS-CoV-2 mRNA vaccination in children aged 12 to 17 years: stratified analysis of a national database https://www.medrxiv.org/content/10.1101/2021.08.30.21262866v1
11. Association of myocarditis with Covid-19 mRNA vaccine in children https://media.jamanetwork.com/news-item/association-of-myocarditis-with-mrna-covid-19-vaccine-in-children/

12. Recurrence of Acute Myocarditis Temporally Associated with Receipt of the mRNA Vaccine in a Male Adolescent https://www.ncbi.nlm.nih.gov/pubmed/34166671
13. Important Insights into Myopericarditis after the Pfizer mRNA Covid-19 Vaccination in Adolescents https://www.ncbi.nlm.nih.gov/pubmed/34332972
14. Covid-19 Vaccination-Associated Myocarditis in Adolescents https://www.ncbi.nlm.nih.gov/pubmed/34389692
15. Recurrence of acute myocarditis temporally associated with receipt of coronavirus mRNA vaccine in a male adolescent https://www.sciencedirect.com/science/article/pii/S002234762100617X
16. Recurrence of myopericarditis following mRNA Covid-19 vaccination in a male adolescent https://www.ncbi.nlm.nih.gov/pubmed/34904134
17. Acute Myocarditis/Pericarditis in Hong Kong Adolescents Following Comirnaty Vaccination https://www.ncbi.nlm.nih.gov/pubmed/34849657
18. Guillain-Barré syndrome after Covid-19 vaccination in an adolescent https://www.pedneur.com/article/S0887-8994(21)00221-6/fulltext
19. **Peri/myocarditis in adolescents after Pfizer-BioNTech Covid-19 vaccine** https://pubmed.ncbi.nlm.nih.gov/34319393/
20. Covid-19 Vaccine for Adolescents. Concern about Myocarditis and Pericarditis https://www.ncbi.nlm.nih.gov/pubmed/34564344
21. Multisystem inflammatory syndrome in a male adolescent after his second Pfizer-BioNTech Covid-19 vaccine https://www.ncbi.nlm.nih.gov/pubmed/34617315
22. Myopericarditis after Pfizer mRNA Covid-19 vaccination in adolescents https://www.sciencedirect.com/science/article/pii/S002234762100665X
23. Myo/pericarditis after vaccination with Covid-19 mRNA in adolescents 12 to 18 years of age https://www.sciencedirect.com/science/article/pii/S0022347621007368
24. Important information on myopericarditis after vaccination with Pfizer Covid-19 mRNA in adolescents https://www.sciencedirect.com/science/article/pii/S0022347621007496
25. Cardiac injury in adolescents receiving mRNA Covid-19 vaccine https://journals.lww.com/pidj/Abstract/9000/Transient_Cardiac_Injury_in_Adolescents_Receiving.95800.aspx
26. Covid-19 vaccine-induced myocarditis: a case report with review of the literature https://www.sciencedirect.com/science/article/pii/S1871402121002253
27. mRNA Coronavirus-19 Vaccine-Associated Myopericarditis in Adolescents: A Survey Study https://www.ncbi.nlm.nih.gov/pubmed/34952008
28. **Myocarditis associated with Covid-19 vaccination in adolescents** https://publications.aap.org/pediatrics/article/148/5/e2021053427/181357

29. **7 cases of acute myocarditis or myopericarditis in healthy male adolescents who presented with chest pain within 4 days after the second dose** https://pubmed.ncbi.nlm.nih.gov/34088762/

30. **Myocarditis and pericarditis in adolescents after first and second doses of Covid-19 mRNA vaccines** https://academic.oup.com/ehjqcco/advance-article/doi/10.1093/ehjqcco/qcab090/64 42104

31. Why are we vaccinating children against Covid-19? https://www.ncbi.nlm.nih.gov/pubmed/34642628

32. Symptomatic Acute Myocarditis in 7 Adolescents after Pfizer-BioNTech Covid-19 Vaccination https://pediatrics.aappublications.org/content/148/3/e2021052478

33. STEMI mimicry: focal myocarditis in an adolescent patient after Covid-19 mRNA vaccination https://pubmed.ncbi.nlm.nih.gov/34756746/

34. **Myocarditis After Covid-19 mRNA Vaccination in Adolescents** https://pubmed.ncbi.nlm.nih.gov/34704459/

35. Myocarditis and pericarditis in adolescents after the first and second doses of Covid-19 mRNA vaccines https://pubmed.ncbi.nlm.nih.gov/34849667/

36. **Cardiovascular magnetic resonance findings in young adult patients with acute myocarditis after Covid-19 mRNA vaccination**: a case series https://pubmed.ncbi.nlm.nih.gov/34496880/

37. Peri/myocarditis after the first dose of mRNA-1273 SARS-CoV-2 (Modern) mRNA-1273 vaccine in a young healthy male: case report https://bmccardiovascdisord.biomedcentral.com/articles/10.1186/s12872-021-02183

38. Perimyocarditis following first dose of the mRNA-1273 SARS-CoV-2 (Moderna) vaccine in a healthy young male https://www.ncbi.nlm.nih.gov/pubmed/34348657

39. Acute myocarditis in a young adult two days after vaccination https://pubmed.ncbi.nlm.nih.gov/34709227/

40. Clinical suspicion of myocarditis temporally related to Covid-19 vaccination in adolescents and young adults: Abnormal findings on cMRI were frequent. Future studies should evaluate risk factors, mechanisms, and long-term outcomes https://pubmed.ncbi.nlm.nih.gov/34865500/

41. Cardiovascular magnetic resonance imaging findings in young adult patients with acute myocarditis after Covid-19 mRNA vaccination: a case series https://jcmr-online.biomedcentral.com/articles/10.1186/s12968-021-00795-4

42. Multimodality imaging and histopathology in a young man presenting with fulminant lymphocytic myocarditis and cardiogenic shock after vaccination with mRNA-1273 https://pubmed.ncbi.nlm.nih.gov/34848416/

43. **Acute myocardial infarction within 24 hours after Covid-19 vaccination** https://pubmed.ncbi.nlm.nih.gov/34364657/

44. Report of a case of myopericarditis after vaccination with BNT162b2 Covid-19 mRNA in a young Korean male https://pubmed.ncbi.nlm.nih.gov/34636504/

45. Clinical Guidance for **Young People with Myocarditis and Pericarditis after Vaccination** with Covid-19 mRNA: There is a temporal association between receiving mRNA Covid-19 vaccination and myocarditis and pericarditis among youth https://www.cps.ca/en/documents/position/clinical-guidance-for-youth-with-myocarditis-and-pericarditis

46. In-depth evaluation of a case of presumed myocarditis after the second dose of Covid-19 mRNA vaccine https://www.ahajournals.org/doi/10.1161/CIRCULATIONAHA.121.056038

47. Acute myocarditis after SARS-CoV-2 vaccination in a 24-year-old male https://pubmed.ncbi.nlm.nih.gov/34334935/

48. Association of myocarditis with the BNT162b2 messenger RNA Covid-19 vaccine in a **case series of children** https://pubmed.ncbi.nlm.nih.gov/34374740/

49. Epidemiology and clinical features of myocarditis/pericarditis before the introduction of Covid-19 mRNA vaccine in Korean children: a multicenter study https://pubmed.ncbi.nlm.nih.gov/34402230/

50. **Acute myopericarditis after Covid-19 vaccination in adolescents** https://pubmed.ncbi.nlm.nih.gov/34589238/

Appendix #2

Sudden Deaths of Healthy Young People

Nearly all incidents in this Appendix were gleaned from local and community news media reports about a healthy young person dying suddenly and unexpectedly. When young people collapse and die at sporting events, the collapses are witnessed by spectators and thus more likely to be reported in news stories. Similarly, in small towns, a young athlete might be well-known to the community, making it more likely that their sudden death will be reported.

In contrast, a child or teenager dying in their sleep in a big city would not likely be reported in the news at all. For these and other reasons, this list makes no effort whatsoever to be a comprehensive collection.

While the number of cases described in this Appendix is large, it is not held out as a complete listing of sudden deaths of healthy young people in 2021 and 2022 — far from it. Trying for a comprehensive collection would fail in any event, given that nearly all of these incidents were ignored by national media companies.

There were two fairly obvious criteria to be met before an incident could be considered for inclusion here:

1. The incident became known beyond the people and families immediately involved (most often by becoming a local news story); or,
2. The incident became known in some other way, such as appearing in the medical or scientific literature.

Though there were just two criteria for inclusion of incidents, there were many criteria for *exclusion*. Rumors, cases merely heard about, and second/third/fourth-hand reports were excluded. Also excluded were deaths about which there was any indication that the person was already severely sick at the time of death (for example, in the care of a doctor or hospital). Also excluded: situations in which there was any indication of suicide, homicide, other foul play, drug overdose, accident, etc.

The news headlines themselves often demonstrate that the deaths were shocking and unexpected. News story photos depicting athletes in team uniforms, for example, make clear that the young person who died was, by all appearances, healthy enough to be included in competitive sports, and that team management was not aware of any health issues for that player, and that dying in that setting and situation is unexpected.

Note: All news headlines and any images included with stories in this book were already public, and newspapers or news outlets had already clearly accomplished their publication goals years in advance of this book, meaning the media companies were not denied any opportunity to publish first. The research team that located information for this book are not aware of any circumstance in which use here could preclude beneficial use of the same material by news outlets, copyright holders, or anyone else. Finally, the team found no instances in which images that had been public in news stories contained herein were objected to or retracted.

Everyone involved in researching and publishing this book expresses their sadness and condolences about these tragic early deaths.

1. **Helena de Marco** **5 yrs** **Died in sleep 8 days post vaccination** Feb 26, 2022
   https://stopcensura.online/bambina-di-5-anni-muore-nel-sonno-8-giorni-dopo-la-prima-dose-pfizer-arresto-cardiaco
2. **Rozalia Spadafora** **5 yrs** **Cardiac event / Myocarditis** Jul 5, 2022
   https://www.abc.net.au/news/2022-07-28/act-family-of-five-year-yrs-girl-died-at-canberra-hospital-speak/101276188
3. **Giulia D.L.** **6 yrs** **Died unexpectedly in her sleep** Apr 1, 2022
   https://edizionecaserta.net/2022/04/27/bimba-di-6-anni-trovata-morta-sul-divano-di-casa-stava-dormendo/
4. **Unnamed Child** **8 yrs** **Died unexpectedly in her sleep** Sep 6, 2022
   https://www.bfmtv.com/police-justice/marmande-un-enfant-de-8-ans-retrouve-mort-a-son-domicile-les-causes-du-deces-totalement-inconnues_AN-202209060305.html
5. **Unnamed Child** **9 yrs** **Sudden death at home in front of parents** Sep 16, 2022
   https://www.fanpage.it/attualita/malore-improvviso-in-casa-si-rivela-fatale-bimbo-di-9-anni-muore-davanti-ai-genitori/
6. **Treven Ball** **10** **Sudden death** Sep 13, 2022
   https://www.kgns.tv/2022/09/15/youth-football-player-dies-unexpectedly-after-fulfilling-dream-playing-under-lights/
7. **Kyran Reading** **10** **Sudden death** Mar 7, 2022
   https://www.peterboroughtoday.co.uk/news/people/kyran-reading-memorial-peterborough-community-gathers-as-football-club-pay-tribute-to-boy-10-who-died-suddenly-3609951
8. **Unnamed Child** **10** **Cardiac event** Apr 19, 2022
   https://www.standard.co.uk/news/london/girl-dies-putney-leisure-centre-cardiac-arrest-unexplained-metropolitan-police-b995087.html
9. **Unnamed Child** **10** **Sudden death during gym class** Aug 2, 2022
   https://www.clarin.com/sociedad/conmocion-tartagal-chico-10-anos-murio-clase-educacion-fisica_0_GhfYdDnYnn.html
10. **Unnamed Child** **11** **Cardiac event** Jul 9, 2022
    https://euroweeklynews.com/2022/07/12/11-year-old-cardiac-arrest-dies-pleneuf-val-andre-france/

11. **Ryan Heffernan 12 Sudden death at school** Mar 28, 2022
https://www.echo-news.co.uk/news/21875632.ryan-heffernan-surplices-family-learning-accept-may-no-answers/

12. **Unnamed Child 12 Sudden death during run** Aug 4, 2022
https://www.nzherald.co.nz/nz/baradene-college-student-death-12-year-old-collapses-during-run/XJVCWWML5MWQL2SSIBDLOM75OU/

13. **Chloe Gavazzi 12 Sudden death** Jun 12, 2021
https://www.notizie.it/cloe-giani-giavazzi-morta-a-milano-a-soli-12-anni-addio-alla-giovanissima-promessa-del-tennis/

14. **Leo Forstenlechner 12 Sudden death** Dec 17, 2021
https://goodsciencing.com/covid/athletes-suffer-cardiac-arrest-die-after-covid-shot/

15. **Carlo Conte 12 Cardiac event during run** Jan 18, 2022
https://www.padovaoggi.it/attualita/carlo-alberto-conte-morto-fiamme-oro-26-gennaio-2022.html

16. **Gemma Caffrey 12 Sudden death** Oct, 2021
https://www.dailyrecord.co.uk/in-your-area/lanarkshire/headteacher-leads-tributes-popular-lanarkshire-25318734

17. **Unnamed Child 12 Cardiac event 1 day post vaccination** Feb 20, 2022
https://au.news.yahoo.com/investigation-into-12-year-old-boys-death-days-after-covid-vaccine-001210961.html

18. **Mattea Sommerville 12 Died unexpectedly in her sleep** Jun, 2022
https://euroweeklynews.com/2022/06/10/12-year-old-girl-dies-suddenly/

19. **Althumani Brown 13 Cardiac event during school field trip** Jun 1, 2022
https://www.cbsnews.com/baltimore/news/baltimore-7th-grader-died-of-natural-causes-on-field-trip-autopsy-finds/

20. **Vanessa Figueiredo 13 Health declined upon vaccination** Jan 10, 2022
https://dailytelegraph.co.nz/covid-19/watch-the-tragic-story-of-vanessa-martins-figueiredo-brazilian-teen-who-died-after-taking-covid-jab/

21. **Chino Nsofor 13 Sudden death during training** Jun 28, 2021
https://www.reviewjournal.com/local/coroner-reveals-cause-of-legacy-high-school-football-players-death-2499882/

22. **Samuel Akwasi 13 Cardiac event during game** May 7, 2022
https://www.mirror.co.uk/sport/football/news/footballer-dies-medical-emergency-nottingham-26905775

23. **Joshua Henry 13 Brain bleed 90 minutes post vaccination** Oct 4, 2021
https://rupreparing.com/news/2021/11/23/joshua-henry-14-year-old-boy-dies-from-massive-brain-bleed-hours-after-receiving-second-pfizer-covid-19-vaccine

24. **Jacob Clynick 13 Cardiac event 3 days post vaccination** Jun 16, 2021
https://nypost.com/2021/07/05/michigan-boy-dies-in-his-sleep-three-days-after-getting-vaccine/

25. **Marco Benitez 13 Sudden death during game** Oct 22, 2021
https://www.kusi.com/memorial-held-for-oceanside-middle-school-student-who-died-suddenly/

26. **Sophia Fenner 13 Sudden death** Jan 17, 2022
https://navarrepress.com/friends-rally-to-help-family-of-13-year-old-girl-who-died-suddenly/

27. **Jack O'Drain 13 Cardiac event** Jan 1, 2022
https://www.gofundme.com/f/the-odrain-family

28. **Jorge Mezcua** **13** **Sudden death** Apr 3, 2022
https://euroweeklynews.com/2022/04/05/young-footballer-died-suddenly-cadiz/

29. **Unnamed Child** **13** **Cardiac event on the playground** Mar 28, 2022
https://euroweeklynews.com/2022/03/28/sudden-death-of-13-year-old-pupil-in-playground-at-the-ies-sierra-de-mijas/

30. **Sheikh Shaheeda** **13** **Cardiac event during class** Sep 7, 2022
http://timesofindia.indiatimes.com/articleshow/94063699.cms

31. **Nico** **14** **Sudden death** Jan, 2022
https://www.merkur.de/lokales/fuerstenfeldbruck/olching-ort29215/bayern-fasching-olching-vorfall-tot-91226280.html

32. **Karol Setniewski** **14** **Sudden death** Dec 17, 2021
https://www.tvp.info/57514609/nie-zyje-mlody-pilkarz-znicza-pruszkow-karol-seta-setniewski-mial-13-lat-byl-kandydatem-do-akademii-legii-warszawa

33. **Wouter Betjes** **14** **Sudden death** Dec 4, 2021
https://www.ad.nl/binnenland/wouter-14-zakt-in-elkaar-op-hockeyveld-en-overlijdt-school-en-club-in-rouw~a064d150/

34. **Talina Rampersad** **14** **Sudden death** Jul 17, 2022
https://www.cbc.ca/news/canada/manitoba/talina-rampersad-husack-death-autopsy-wait-1.6534329

35. **Giada Furlanut** **14** **Blood clot** Dec 5, 2021
https://pledgetimes.com/giada-dies-at-the-age-of-14-after-having-been-ill-at-school/

36. **Aoibhe Byrne** **14** **Sudden death** Apr 27, 2022
https://www.irishmirror.ie/news/irish-news/monaghan-community-state-shock-following-26824075

37. **Unnamed Child** **14** **Cardiac event during training** Sep 7, 2022
https://www.ansa.it/english/news/general_news/2022/09/07/boy-14-dies-5-days-after-cardiac-arrest-on-soccer-pitch_f602d1d8-1d35-4c6a-976d-1e30fc8c0548.html

38. **Milagros Santiso** **14** **Cardiac event** Jul 30, 2022
https://aleteia.org/2022/08/20/14-year-old-girl-dies-on-a-retreat-her-mothers-words-of-faith-will-inspire-you/

39. **Unnamed Child** **14** **Sudden death during game** Aug 4, 2021
https://www.insidehalton.com/news-story/10450254-milton-boy-14-dies-after-collapsing-in-parking-lot-while-playing-basketball/

40. **Jack Pollock** **14** **Sudden death** Sep 29, 2022
https://www.edinburghlive.co.uk/news/edinburgh-news/tributes-loving-edinburgh-schoolboy-who-25159335

41. **Unnamed Child** **15** **Cardiac event during training** Apr 29, 2022
https://www.francebleu.fr/infos/faits-divers-justice/un-adolescent-de-15-meurt-d-un-malaise-cardiaque-pendant-un-entrainement-de-football-a-contrexeville-1651307095

42. **Unnamed Child** **15** **Sudden death** Aug 3, 2022
https://www.mmegi.bw/sports/mexican-girls-starlet-collapses-dies-in-training/news

43. **DeVonte Mumphrey** **15** **Sudden death during game** Feb 8, 2022
https://www.newsweek.com/devonte-mumphrey-texas-high-school-basketball-game-alto-texas-1677533

44. **Brian Saes** **15** **Cardiac event while cycling** Jul 21, 2021
https://g1.globo.com/sc/santa-catarina/noticia/2021/07/22/adolescente-morre-durante-passeio-de-bicicleta-em-sc-era-um-menino-muito-feliz-diz-tia.ghtml

45. **Isaiah Banks** **15** **Sudden death** Jul 11, 2021
https://www.fox5atlanta.com/news/norcross-high-school-mourns-football-player-isaiah-banks-death

46. **Stephen Sylvester** **15** **Sudden death** Aug 2, 2021
https://www.theoaklandpress.com/2021/08/10/catholic-central-mourning-the-loss-of-football-track-athlete-stephen-sylvester/

47. **Pedro Oliveira** **15** **Sudden death** Jan 11, 2022
https://www.thegatewaypundit.com/2022/01/15-year-old-soccer-player-brazil-dies-cardiac-arrest-following-national-football-cup-tournament-game/

48. **Preston Settles** **15** **Cardiac event during game** Feb 5, 2022
https://www.cbsnews.com/boston/news/preston-settles-funeral/

49. **Elias Georgakopoulos** **15** **Sudden death 3 days post vaccination** Oct, 2021
https://www.eventiavversinews.it/elias-georgakopoulos-15-anni-perfettamente-sano-muore-3-giorni-dopo-aver-ricevuto-il-vaccino-pfizer-covid-19-il-fratello-racconta/

50. **Danylo Nobre** **15** **Became sick 18 days post vaccination** Mar 3, 2022
https://community.covidvaccineinjuries.com/danylo-15-year-old-died-after-pfizer-induced-brainstem-encephalitis/

51. **Asheley Garcia** **15** **Stroke 5 days post vaccination** Mar 13, 2022
https://diariodevallarta.com/en/la-muerte-de-asheley-carbajal-garcia-si-no-se-hubiera-vacunado-estaria-viva/

52. **Matteo Pietrosanti** **15** **Sudden death during training** Mar 3, 2022
https://pledgetimes.com/latina-matteo-pietrosanti-dies-at-the-age-of-15-in-front-of-his-mothers-eyes/

53. **Unnamed Child** **15** **Sudden death 2 days post vaccination** Jun 7, 2021
https://www.pressdemocrat.com/article/news/county-officials-social-media-posters-spar-over-boys-death/

54. **Nico Holguin** **15** **Sudden death** Mar 21, 2022
https://www.infobierzo.com/fallece-en-leon-el-joven-nico-holguin-de-15-anos-que-fue-reanimado-en-el-parque-del-plantio-de-ponferrada/669304/

55. **Carmyne Payton** **15** **Sudden death during game** Nov 18, 2021
https://people.com/sports/15-year-old-boy-collapses-dies-basketball-tryouts-long-island-new-york/

56. **Unnamed Child** **16** **Cardiac event during game** Feb 25, 2022
https://primeraplana.mx/archivos/858300

57. **Unnamed Child** **16** **Sudden death during training** Sep 6, 2022
https://www.cbsnews.com/baltimore/news/randallstown-high-school-student-dies-after-medical-emergency-at-football-practice/?intcid=CNM-00-10abd1h

58. **Anna Burns** **16** **Cardiac event at cross-country meet** Sep 13, 2022
https://www.amherstbulletin.com/Anna-Burns-remembered-by-coaches-teammates-48204979

59. **Henry Farmer** **16** **Sudden death** Apr 12, 2022
https://patch.com/connecticut/darien/darien-high-hockey-player-remembered-great-teammate-hard-worker

60. **Ernesto Ramirez, Jr.** **16** **Cardiac event 5 days post vaccination** Apr 24, 2021
https://circleofmamas.com/health-news/grieving-father-ernest-ramirez-shares-heartbreaking-story-of-his-teen-sons-death-5-days-after-pfizer-vaccine/

61. **Joshua Johnson** **16** **Sudden death** Jun, 2021
https://dailyprogress.com/community/greenenews/news/memorial-to-remember-joshua-johnson-on-thursday-june-10/article_98a9194c-c942-11eb-b70d-d3d85f6333a1.html

62. **Ethan Trejo** **16** **Sudden death** Jun 25, 2021
https://www.cincinnati.com/story/news/2021/06/25/teen-dies-after-medical-incident-princeton-high-school-field/5344293001/

63. **Devon DuHart** **16** **Sudden death** Jul 24, 2021
https://www.thv11.com/article/sports/little-rock-central-death-devon-duhart/91-82359c68-9a8e-4611-90fb-cf679ce1ee72

64. **Jascha Zey** **16** **Sudden death** Jul 28, 2021
https://www.sportfreunde-eisbachtal.de/ploetzlich-und-viel-zu-frueh-die-eisbaeren-familie-trauert-um-u19-spieler-jascha-zey

65. **Nathan Esparza** **16** **Cardiac event** Jul 13, 2021
https://scvnews.com/castaic-high-school-brings-grief-counselors-on-campus-after-student-death/

66. **Jamarcus Hall** **16** **Sudden death** Aug 11, 2021
https://www.wlbt.com/2021/11/09/16-year-old-mississippi-football-player-dies/

67. **Antonio Hicks** **16** **Sudden death** Sep 28, 2021
https://www.nfldraftdiamonds.com/2021/10/antonio-hicks/

68. **Jony Lopez** **16** **Cardiac event** Nov 11, 2021
https://radioconcierto.com.py/2021/11/12/futbolista-infarto-durante-practica/

69. **Valentin Rodionov** **16** **Sudden death during game** Nov 28, 2021
https://www.rt.com/sport/541525-russian-ice-hockey-young-star-dies/?s=09

70. **Mosheur Rahman** **16** **Adverse events upon vaccination** Aug 30, 2021
https://fleekus.com/b/antonio-silva-fkzle/article/mosheur-rahman-healthy-16-year-old-boy-dies-shortly-after-receiving-the-moderna-covid-19-vaccine-family-seeks-justice-fgtyq

71. **Melanie Macip** **16** **Cardiac event 5 days post vaccination** Aug 7, 2021
https://twitter.com/veritebeaute/status/1439363821370494977

72. **Sofia Benharira** **16** **Cardiac event / Blood clot** Sep 21, 2021
https://www.australiannationalreview.com/covid-19-deaths-and-injuries/sofia-benharira-16-years-old-died-from-pfizer-vaxxine/

73. **Isabelli Valentim** **16** **Sudden death** Sep 2, 2021
https://thecovidblog.com/2021/09/23/isabelli-borges-valentim-16-year-old-brazilian-girl-develops-blood-clots-dead-eight-days-after-first-pfizer-mrna-injection/

74. **Kamrynn Thomas** **16** **Cardiac event soon post vaccination** Mar 30, 2021
https://healthimpactnews.com/2021/16-year-old-wisconsin-girl-dead-following-2-doses-of-the-experimental-pfizer-covid-injections/

75. **Amy Forde** **16** **Died unexpectedly in her sleep** Nov 23, 2021
https://www.mylondon.news/news/east-london-news/healthy-girl-16-died-out-24349492

76. **Lee Nolan** **16** **Sudden death** May 13, 2022
https://extra.ie/2022/05/19/news/irish-news/community-in-wexford-deeply-saddened-following-death-of-incredibly-kind-young-student

77. **Kevin Amaya** **16** **Cardiac event during game** Jul 10, 2022
https://www.periodicoequilibrium.com/adolescente-de-16-anos-muere-de-un-infarto-durante-partido-de-futbol/

78. **Unnamed Child** **16** **Sudden death during training** Jun 8, 2022
https://www.wtkr.com/news/bayside-high-school-student-athlete-dies-after-collapsing-during-conditioning

79. **Owen Cotty** **16** **Cardiac event while playing Frisbee** Aug 6, 2022
https://patch.com/pennsylvania/lowerprovidence/montco-teen-dies-cardiac-arrest-while-playing-frisbee

80. **Gianluca Schettino** **16** **Sudden death** May 15, 2022
https://napoli.occhionotizie.it/gragnano-ragazzo-morto-gianluca-funerali-16-anni/

81. **Quentin Watson** **16** **Sudden death** Aug 18, 2022
https://dailyvoice.com/pennsylvania/montgomery/news/beloved-norristown-area-high-school-student-dies-suddenly-at-16/841307/

82. **D.J.** **16** **Sudden death during game** Sep 19, 2021
https://www.telegraf.rs/english/3393101-young-football-players-pulse-returned-briefly-but-he-couldnt-be-saved-and-passed-away-in-seconds

83. **Cameran Wheatley** **17** **Sudden death during game** Feb 8, 2022
https://abc7chicago.com/cameran-wheatley-bremen-high-school-basketball-christ-hospital/11549978/

84. **Bailey Munro** **17** **Sudden death** Jul 21, 2022
https://www.pressandjournal.co.uk/fp/news/inverness/4576309/death-of-17-year-old-in-inverness-bailey-matheson-munro/

85. **Ali Muhammad** **17** **Died unexpectedly in his sleep** Sep 8, 2022
https://www.westernjournal.com/seemingly-healthy-17-year-old-football-player-dies-sleep-will-forever-thoughts/

86. **Shruti Soni** **17** **Cardiac event** Sep 30, 2021
https://www.freepressjournal.in/bhopal/madhya-pradesh-teen-players-death-due-to-cardiac-arrest-triggers-concern

87. **Miguel Lugo** **17** **Sudden death during training** Mar 3, 2021
https://www.recordonline.com/story/sports/high-school/2021/03/02/wallkill-football-player-miguel-lugo-dies-after-practice-on-monday/6894023002/

88. **Andrew Roseman** **17** **Sudden death** Jul 15, 2021
https://www.phillyvoice.com/red-land-baseball-player-dies-andrew-roseman-pennsylvania-york-county/

89. **Donadrian Robinson** **17** **Sudden death** Aug 29, 2021
https://www.wistv.com/2021/09/04/he-would-love-it-donadrian-robinsons-family-reacts-tribute-wj-keenan-high-school/

90. **Sean Hartman** **17** **Myocarditis 4 days post vaccination** Sep 27, 2021
https://www.lifesitenews.com/news/all-he-wanted-to-do-was-play-hockey-grieving-dad-says-pfizer-shot-killed-his-17-year-old-son/

91. **Dylan Rich** **17** **Cardiac event during game** Sep 7, 2021
https://www.bbc.com/news/uk-england-nottinghamshire-58462925

92. **Krystian Kozek** **17** **Sudden death** Oct 31, 2021
https://gol24.pl/nie-zyje-krystian-kozek-17letni-zawodnik-wisloka-strzyzow/ar/c2-15879099

93. **Nathan Rogalski** **17** **Sudden death** Jan 23, 2022
https://www.oklahoman.com/story/sports/high-school/baseball/2022/01/23/deer-creek-high-school-baseball-player-dies-sudden-illness-nathan-rogalski/6632572001/

94. **Viggo Sorensen** **17** **Cardiac event** Jan 27, 2022
https://www.facebook.com/GEMSWellingtonAcademy.AlKhail.Dubai/photos/a.215399978658681/1784485951750068/?type=3

95. **Philip Laster Jr.** **17** **Sudden death during training** Aug 1, 2022
https://www.wapt.com/article/brandon-high-school-football-player-dies-district-confirms/40777412#

96. **Cesar Vasquez** **17** **Sudden death** Aug 2, 2022
https://www.azcentral.com/story/sports/high-school/2022/08/04/peoria-centennial-football-community-rocked-death-player/10234394002/

97. **Shubham Chopde** **17** **Cardiac event** Jul 29, 2022
https://indianexpress.com/article/cities/pune/pune-student-dies-of-suspected-cardiac-arrest-on-college-trip-to-raireshwar-fort-8060438/

98. **Gwen Casten** **17** **Died unexpectedly in her sleep** Jun 13, 2022
https://www.cbsnews.com/chicago/news/gwen-casten-congressman-sean-casten-daughter-death-sudden-cardiac-arrhythmia-heart-condition/

99. **Adam Ali** **17** **Sudden death** Sep, 2021
https://www.birminghammail.co.uk/news/midlands-news/a-truly-gentle-soul-tributes-21763398

100. **Unnamed Child** **17** **Sudden death** Sep, 2022
https://www.leggo.it/italia/milano/morto_casa_17_anni_varazze_cosa_successo-6908139.html

101. **Tyler Erickson** **17** **Sudden death during training** Sep 12, 2022
https://www.wjhg.com/2022/09/13/community-mourns-death-holmes-county-athlete/

102. **Ali Muhamad** **17** **Died unexpectedly in his sleep** Sep 8, 2022
https://unionnewsdaily.com/sports/rahway-varsity-football-player-dies-in-his-sleep-team-to-honor-him-this-season

103. **Kooper McCabe** **17** **Sudden death** Aug 30, 2022
https://www.mansfieldnewsjournal.com/story/news/2022/09/02/galion-student-athlete-remembered-after-his-unexpected-death/65466123007/

104. **Rohan Cosgriff** **17** **Sudden death** Jul 29, 2022
https://www.racing.com/news/2022-07-31/news-industry-ballarat-mourns-cosgriff

105. **Alessia De Nadai** **17** **Sudden death** Jul 4, 2022
https://www.ilparagone.it/cronaca/alessia-muore-a-17-anni-il-malore-i-due-interventi-alla-testa-e-la-fine-origine-sconosciuta/

106. **Paola Alcantara** **17** **Cardiac event during lunch break** May 26, 2022
https://euroweeklynews.com/2022/05/27/17-year-old-girl-cardiac-arrest/

107. **Eitan Force** **17** **Sudden death during game** Sep 21, 2022
https://www.msn.com/en-us/news/us/high-school-student-dies-during-flag-football-game/ar-AA127NPi?li=BBnb7Kz

108. **Blake Barklage** **17** **Cardiac event during game** Oct 30, 2021
https://6abc.com/blake-barklage-death-lasalle-college-high-school-montgomery-county/11187002/

109. **Estefania Arroyo** **18** **Sudden death** Sep 25, 2021
https://www.cuestonian.com/cuesta-college-student-athlete-cause-of-death-remains-a-mystery/

110. **Adrien Sandjo** **18** **Cardiac event during game** Dec 22, 2021
https://sportnewsafrica.com/en/at-a-glance/italy-adrien-sandjo-cameroonian-footballer-dies-after-a-cardiac-arrest/

111. **Alberto Torrecilla** **18** **Sudden death** Jan 20, 2022
https://www.eventiavversinews.it/madrid-muore-per-arresto-cardiaco-improvviso-giovedi-20-gennaio-il-calciatore-alberto-torrecilla-delle-giovanili-del-club-deportivo-avance/

112. **Kasey Turner** **18** **Blood clot 14 days post vaccination** Feb 27, 2021
https://www.yorkshirepost.co.uk/health/family-of-teenager-who-died-after-having-astrazeneca-jab-pay-tribute-to-cheeky-daughter-3625541

113. **Jacob Downey** **18** **Cardiac event** Sep 29, 2021
https://www.thepeterboroughexaminer.com/sports/hockey/2021/10/01/a-brilliant-kid-on-and-off-the-ice-and-in-every-sport-he-played.html

114. **Emmanuel Antwi** **18** **Sudden death during game** Mar 22, 2021
https://www.cbsnews.com/sacramento/news/kennedy-high-emmanuel-antwi-collapses-during-game-dies/

115. **Nikita Sidorov** **18** **Sudden death during game** Apr 4, 2021
https://www.rt.com/sport/520142-russian-football-player-death-znamya-truda/

116. **Victor Hegedus** **18** **Sudden death during training** Jun 21, 2021
https://www.budapestherald.hu/sport/2021/06/26/an-18-year-old-hungarian-football-player-collapsed-and-died-during-training/

117. **Jack Alkhatib** **18** **Cardiac event during training** Aug 24, 2021
https://www.wistv.com/2021/08/28/dutch-fork-high-school-honors-life-jack-alkhatib-with-memorial/

118. **Amir Aiana** **18** **Cardiac event during game** Jan 11, 2022
https://www.milanotoday.it/cronaca/morto-oratorio-san-giustino.html

119. **Mateo Hernandez** **18** **Sudden death** Jan 11, 2022
https://www.eventiavversinews.it/muore-martedi-11-gennaio-a-18-anni-per-malore-improvviso-il-portiere-spagnolo-mateo-hernandez/

120. **Lucas Dias** **18** **Sudden death during workout** Jan 15, 2022
https://www.tribunadecianorte.com.br/ultimas-noticias/cianorte-policia-aguarda-laudo-para-definir-causa-da-morte-de-jovem-em-academia/

121. **Leo McBride** **18** **Sudden death** Oct 24, 2021
https://www.irishmirror.ie/news/irish-news/community-pays-tribute-teenager-heart-25294844

122. **Alessia Raiciu** **18** **Died unexpectedly in her sleep** Aug 15, 2022
https://tinyurl.com/5xnvj5ex

123. **Valentina Yazenok** **18** **Neurological events days post vaccination** Jan 11, 2022
https://kam24.ru/news/main/20220112/86541.html

124. **Dani Gómez** **18** **Sudden death** Aug 23, 2022
https://aragondigital.es/deportes/2022/08/23/fallece-dani-gomez-jugador-de-18-anos-del-penas-huesca-de-baloncesto/

125. **Filippo Venezia** **18** **Sudden death** Aug 19, 2022
https://www.ilmessaggero.it/sport/altrisport/filippo_dalla_venezia_morto_casa_rugby_mogliano-6878936.html

126. **Avery Gilbert** **18** **Sudden death** Aug 10, 2022
https://patch.com/illinois/grayslake/s/id0uy/football-player-dies-after-collapsing-on-college-campus-in-deerfield

127. **Jessica Matthews** **18** **Sudden death during game** Jun 15, 2022
https://www.dailymaverick.co.za/article/2022-06-17-western-province-and-maties-hockey-player-collapses-dies-on-field/

128. **Alejandro Candela** **18** **Sudden death** Jun 14, 2022
https://www.lavozdelanzarote.com/actualidad/sociedad/fallece-repentinamente-el-joven-alejandro-candela-una-de-las-promesas-de-la-natacion-en-lanzarote_128553_102_amp.html

129. **Tobiloba Taiwo** **18** **Sudden death during game** Feb 21, 2022
https://mndaily.com/271256/news/breaking-umn-student-dies-unexpectedly-at-recwell-center/

130. **Camilla Canepa** **18** **Blood clot days post vaccination** Jun 10, 2021
https://genova.repubblica.it/cronaca/2021/06/10/news/e_morta_camilla_canepa_la_18enne_ligure_vaccinata_con_astrazeneca-305366834/

131. **Davide Bristot** **18** **Died in sleep weeks post vaccination** Jul 14, 2021
https://www.true-news.it/facts/davide-bristot-morto-a-18-anni-perche-si-indaga-sul-vaccino

132. **Kacper Zabrzycki** **18** **Sudden death** Aug 22, 2021
https://www.o2.pl/sport/nie-zyje-kacper-zabrzycki-mial-tylko-18-lat-jestesmy-wstrzasnieci-6675225822878368a

133. **Aidan Kaminska** **19** **Sudden death** May 30, 2022
https://www.the-sun.com/news/5475432/aidan-kaminska-umass-lacrosse-player-died/

134. **Unnamed Person** **19** **Cardiac event during run** Sep 6, 2022
https://www.linfokwezi.fr/un-gamin-de-19-ans-emporte-par-une-crise-cardiaque-pendant-un-footing/

135. **Keanu Breurs** **19** **Sudden death during training** Jan 12, 2021
https://www.hln.be/beveren/dinsdag-nog-op-training-woensdag-plots-overleden-voetbalclub-svelta-in-rouw-na-verlies-van-talentvolle-en-betrokken-speler-keanu-19~a7e8b759/

136. **Kamila Label-Farrell** **19** **Sudden death during run** Jun 9, 2021
https://www.baytoday.ca/obituaries/lebel-farrell-kamila-3884874

137. **Aidan Price** **19** **Sudden death** Jun 20, 2021
https://www.carleton.edu/farewells/aidan-price-24/

138. **Whitnee Abriska** **19** **Cardiac event** Jul 26, 2021
https://www.dhnet.be/sports/sport-regional/liege/2021/07/26/la-joueuse-du-femina-vise-decede-subitement-a-lage-de-19-ans-PV6FI7R2QNBZTMUTUNNINTPXBY/

139. **Marco Tampwo** **19** **Cardiac event** Aug 15, 2021
https://www.thesun.co.uk/sport/football/15892036/footballer-marco-tampwo-dead-heart-attack-covid/

140. **Tirrell Williams** **19** **Stroke during training** Aug 4, 2021
https://www.fourstateshomepage.com/news/local-news/fscc-football-player-passes-away/

141. **Quandarius Wilburn** **19** **Sudden death during training** Aug 8, 2021
https://www.nbcnews.com/news/us-news/virginia-union-university-football-player-dies-after-collapsing-during-practice-n1276410

142. **Sebastiaan Bos** **19** **Sudden death** Sep 11, 2021
https://goodsciencing.com/covid/athletes-suffer-cardiac-arrest-die-after-covid-shot/

143. **Anna Biktimirova** **19** **Sudden death** Dec 1, 2021
https://www.rt.com/sport/541846-arina-biktimirova-taekwondo-perm-death/

144. **Zachary Icenogle** **19** **Died unexpectedly in his sleep** Dec 26, 2021
https://patch.com/illinois/plainfield/long-live-ice-teens-unexpected-death-has-community-mourning

145. **Volodymyr Salo** **19** **Cardiac event hours post vaccination** Sep 13, 2021
https://expose-news.com/2021/09/29/19-year-old-ukrainian-student-gets-pfizer-vaccine-behind-his-familys-back-dies-seven-hours-later/

146. **Inês Rafael** **19** **Sudden death 5 days post vaccination** Aug, 2021
https://www.cmjornal.pt/portugal/detalhe/universitaria-morre-cinco-dias-apos-a-vacina-da-covid-19-em-vieira-do-minho?ref=HP_PrimeirosDestaques

147. **Simone Scott** **19** **Cardiac event / Myocarditis** Jun 11, 2021
https://www.fox19.com/2021/06/16/mason-high-school-graduate-remembered-kind-talented-after-mysterious-illness-takes-her-life/

148. **Jaysley-Louise Beck** **19** **Sudden death** Dec 15, 2021
https://www.wiltshirelive.co.uk/news/wiltshire-news/familys-heart-breaking-tribute-beautiful-6534286

149. **Matthias Pedersen** **19** **Sudden death** Apr 24, 2022
https://sport.tv2.dk/haandbold/2022-04-24-u20-landsholdsspiller-matthias-birkkjaer-er-pludselig-doed

150. **Teun Elbers** **19** **Sudden death while hiking** Aug 5, 2022
https://www.telegraaf.nl/nieuws/908171433/voetbalclub-oss-rouwt-om-overleden-teun-elbers-19

151. **Stephanie Ming** **19** **Stroke** Jun 19, 2022
https://www.theborneopost.com/2022/06/23/sarawak-sukma-shooter-stephanie-dies-at-19/

152. **Giuseppe Gallina** **19** **Sudden death during game** Feb 12, 2022
https://41esimoparallelo.it/2022/02/22/tragedia-a-statte-e-giuseppe-gallina-il-19enne-morto-durante-la-partita-di-calcetto-sotto-gli-occhi-dei-compagni-aperta-uninchiesta/37/

153. **Rittika Das** **19** **Cardiac event during workout** Aug 9, 2022
https://timesofindia.indiatimes.com/city/kolkata/19-year-old-falls-ill-at-gym-dies-in-kolkata/articleshow/93465703.cms

154. **Brazil Walsh** **20** **Sudden death** Apr 24, 2022
https://www.leeds-live.co.uk/news/leeds-news/woman-20-died-rare-brain-24213876

155. **Derek Gray** **20** **Cardiac event during game** Jul 24, 2022
https://www.msn.com/en-us/sports/ncaabk/college-basketball-star-derek-gray-dead-at-20/ar-AA1072PN

156. **Lily Ernst** **20** **Sudden death** Jul 27, 2022
https://unipanthers.com/news/2022/7/28/womens-swimming-and-diving-uni-mourns-the-loss-of-swimming-student-athlete-lily-ernst.aspx

157. **Eli Palfreyman** **20** **Sudden death during half-time** Aug 30, 2022
https://kitchener.ctvnews.ca/ayr-centennials-missing-the-heart-and-soul-of-our-team-after-captain-s-death-1.6052332

158. **Christian Blandini** **20** **Cardiac event** Sep 9, 2021
https://freewestmedia.com/2021/09/16/sudden-death-of-young-italian-athlete-and-the-conspiracy-of-silence/

159. **Kim Kyeong-Bo** **20** **Sudden death** Nov 8, 2021
https://www.gamespot.com/articles/professional-overwatch-player-kim-alarm-kyeong-bo-dies-at-20/1100-6497812/

160. **Vinicius Freitas** **20** **Cardiac event** Dec 3, 2021
https://www.meiahora.com.br/esportes/2021/12/6290244-jovem-com-passagem-por-clube-carioca-morre-de-infarto-aos-20-anos.html

161. **Ali Arabzada** **20** **Sudden death** Dec 17, 2021
https://tolonews.com/sport-175926

162. **Tatjana Jagodic** **20** **Blood clot / Brain bleed** Sep, 2021
https://slovenia.postsen.com/local/28539/It-will-be-a-year-since-the-death-of-Katja-for-whom-the-vaccine-against-covid-19-was-fatal.html

163. **Regan Lewis** **20** **Cardiac event 1 day post vaccination** Sep 27, 2022
https://citizenfreepress.com/breaking/healthy-young-student-is-dead-one-day-after-covid-vaccine/

164. **Djouby Laura** **20** **Cardiac event** Aug 11, 2022
https://monewsguyane.com/2022/08/12/un-footballeur-de-lusc-roura-meurt-dun-arret-cardiaque/

165. **Andrea Musiu** **20** **Sudden death** Jul 23, 2022
https://www.ilmessaggero.it/italia/andrea_musiu_morto_calcetto_partita_cagliari_chi_era_ultime_notizie-6831104.html

166. **Unnamed Person** **20** **Cardiac event during marathon** May 22, 2022
https://www.waz.de/staedte/gelsenkirchen/zusammenbruch-vor-ziel-mann-stirbt-bei-vivawest-marathon-id235422969.html

167. **Oliver Vaux** **20** **Died unexpectedly in his sleep** May 26, 2022
https://www.heraldscotland.com/news/20202278.university-pays-tribute-exemplary-student-oliver-vaux-dies-sleep/

168. **Andres Melendez** **20** **Sudden death** Dec 16, 2021
https://www.cleveland.com/guardians/2022/01/what-caused-the-death-of-cleveland-guardians-minor-leaguer-andres-melendez-hey-hoynsie.html

169. **Joanna Krudys** **21** **Sudden death post vaccination** Dec 4, 2021
https://nczas.com/2021/12/09/nagla-smierc-dwojki-wroclawskich-studentow-internet-wrze-foto/

170. **Awysum Harris** **21** **Sudden death** Jul 3, 2022
https://www.waff.com/2022/07/05/alabama-state-mourns-death-football-player/

171. **Reda Saki** **21** **Sudden death during game** Apr 11, 2021
https://www.moroccoworldnews.com/2021/04/339555/21-year-old-moroccan-football-player-dies-after-collapsing-on-pitch

172. **Alexey Zelenin** **21** **Cardiac event during training** May 12, 2022
https://markcrispinmiller.substack.com/p/in-memory-of-those-who-died-suddenly-a59?s=r

173. **Fabricio Navarro** **21** **Died unexpectedly in his sleep** Jun 15, 2022
https://442.perfil.com/noticias/futbol/dolor-en-atletico-tucuman-por-el-fallecimiento-de-uno-de-sus-jugadores.phtml

174. **Marvel Simiyu** **21** **Sudden death** Jun 14, 2022
https://euroweeklynews.com/2022/06/16/young-female-footballer-marvel-simiyu-dies-suddenly/

175. **Aliya Khambikova** **21** **Sudden death** Nov 7, 2021
https://www.rt.com/sport/539670-russian-volleyball-death-aliya-khambikova/

176. **Nelson Solano** **21** **Cardiac event** Nov 8, 2021
https://www.abc.com.py/nacionales/2021/11/07/joven-futbolista-fallece-de-un-infarto-despues-de-un-partido/

	Name	Age	Event	Date
177.	**Dawid Akula**	**21**	**Sudden death during game**	Dec 4, 2021
	https://nczas.com/2021/12/09/nagla-smierc-dwojki-wroclawskich-studentow-internet-wrze-foto/			
178.	**Aurelie Hans**	**21**	**Cardiac event**	Dec 14, 2021
	https://www.dna.fr/societe/2021/12/18/apres-le-deces-d-aurelie-hans-l-emotion-dans-le-monde-du-football-feminin			
179.	**Bryce Murray**	**21**	**Sudden death**	Dec 27, 2021
	https://www.thescottishsun.co.uk/news/scottish-news/8273835/forfar-scaffolder-suden-death-tributes/			
180.	**Herbert Afayo**	**21**	**Cardiac event during training**	Jan 2022
	https://theinvestigatornews.com/2022/01/oh-no-the-sad-story-of-how-footballer-hebert-afayo-collapsed-and-instantly-died-on-pitch/			
181.	**Alexandros Lampis**	**21**	**Cardiac event during game**	Feb 2, 2022
	https://www.thesun.co.uk/sport/17524888/alexandros-lampis-dies-21-greek-footballer-cardiac-arrest/			
182.	**Sanjay Vimalraj**	**21**	**Sudden death**	Jul, 2022
	https://timesofindia.indiatimes.com/city/chennai/tamil-nadu-cm-announces-rs-3lakh-solatium-to-kin-of-deceased-kabaddi-player/articleshow/93166087.cms			
183.	**Clark Yarbrough**	**21**	**Sudden death**	Sep 4, 2022
	https://www.si.com/college/2022/09/05/ouachita-baptist-defensive-lineman-clark-yarbrough-dies-at-21			
184.	**Renjitha**	**21**	**Blood clot days post vaccination**	Aug, 2021
	https://keralakaumudi.com/en/news/news.php?id=625121&u=%20news.php?id=21-year-old-dies-in-kasargod-days-after-taking-covid-jab			
185.	**John Foley**	**21**	**Died in sleep hours post vaccination**	Apr 11, 2021
	https://thecovidblog.com/2021/04/14/john-francis-foley-21-year-old-university-of-cincinnati-student-dead-24-hours-after-johnson-johnson-shot/			
186.	**Dominyka Podziute**	**21**	**Sudden death**	Apr 13, 2022
	https://www.mirror.co.uk/sport/football/news/dominyka-podziute-dead-former-newcastle-26710872			
187.	**Laura Domecq**	**21**	**Sudden death**	Aug 7, 2022
	https://www.monacomatin.mc/football/une-marche-organisee-ce-mercredi-soir-en-hommage-a-une-jeune-joueuse-de-las-monaco-decedee-791963			
188.	**Gianmarco Verdi**	**21**	**Sudden death during dinner**	May 28, 2022
	https://www.ilrestodelcarlino.it/modena/cronaca/gianmarco-verdi-1.7726078			
189.	**Justin Tabone**	**22**	**Cardiac event during game**	Oct 10, 2022
	https://timesofmalta.com/articles/view/22yearold-dies-collapsing-football-pitch.986852			
190.	**Hayden Holman**	**22**	**Cardiac event during marathon**	Oct 4, 2021
	https://www.stgeorgeutah.com/news/archive/2022/09/29/ggg-friends-family-will-commemorate-hayden-holmans-life-at-46th-running-of-st-george-marathon/#.YO3wkC-B1-U			
191.	**Imtiyaz Khan**	**22**	**Cardiac event during game**	Sep 2, 2022
	https://www.greaterkashmir.com/kashmir/pulwama-youth-dies-of-suspected-heart-attack-while-playing-cricket-in-anantnag			
192.	**Bruno Macedo**	**22**	**Sudden death**	Nov 21, 2021
	https://www.ouest-france.fr/nouvelle-aquitaine/nueil-les-aubiers-79250/faits-divers-deces-d-un-jeune-homme-de-22-ans-il-jouait-au-fc-nueil-les-aubiers-122cc602-4af2-11ec-8a6b-582d17cbe42b			

193. **Dejmi Dumervil** **22** **Sudden death** Nov 12, 2021
https://sports.yahoo.com/former-louisville-football-player-dejimi-154847843.html

194. **Fatimah Shabazz** **22** **Sudden death** Nov 30, 2021
https://greensboro.com/sports/college/a-t-volleyball-player-fatimah-shabazz-dies-suddenly/article_8c531018-521c-11ec-9cde-fb75a01ce59d.html

195. **Patricio Guaita** **22** **Sudden death during training** Jan 18, 2022
https://www.ole.com.ar/informacion-general/fallecimiento-patricio-guaita-jugador-22-anos-plata_0_jvyHHTY7b.html

196. **Michael Almanza** **22** **Cardiac event during game** Mar 20, 2022
https://euroweeklynews.com/2022/03/22/young-footballer-heart-attack/

197. **Arthur Grice** **22** **Neurological events day post vaccination** Feb 24, 2022
https://circleofmamas.com/health-news/22-year-old-arthur-grice-died-6-weeks-after-johnson-johnson-vaccine-from-paralytic-ileus/

198. **Mubarak Sayed** **22** **Cardiac event** Jul, 2022
https://www.mirror.co.uk/news/world-news/university-student-dies-joy-after-27613873

199. **Costa Debochado** **22** **Sudden death** Jan 22, 2021
https://www.breaktudo.com/tiktoker-leo-costa-debochado-morre-aos-18-anos-de-idade-apos-passar-por-problemas-pulmonares/

200. **Maxime Beltra** **22** **Sudden death 9 hours post vaccination** Jul 26, 2021
https://thecovidblog.com/2021/08/03/maxime-beltra-22-year-old-french-man-dead-nine-hours-after-first-experimental-pfizer-mrna-injection/

201. **Davis Heller** **22** **Sudden death** Oct 6, 2022
https://www.usatoday.com/story/sports/2022/10/07/alabama-transfer-ngu-baseball-player-davis-heller-dies-at-age-22-north-greenville/69546077007/

202. **Andres Burgos** **22** **Cardiac event** Jun 27, 2022
https://www.molinabasket.es/es/publication/126756

203. **Georgia Solanaki** **22** **Cardiac event during training** Jun 15, 2022
https://www.newsy-today.com/tragedy-during-training-the-22-year-old-died/

204. **Samuel Carletti** **22** **Cardiac event** Mar 22, 2022
https://www.trevisotoday.it/cronaca/selva-volpago-montello-samuel-carletti-22-marzo-2022.html

205. **Unnamed Person** **22** **Cardiac event during game** Mar 23, 2022
https://www.telenoche.com.uy/sociedad/fallecio-un-infarto-jugador-amateur-division-40-n5326902

206. **Ben Penrose** **22** **Cardiac event** Aug 16, 2022
https://www.theleader.com.au/story/7870298/fundraiser-for-young-man-with-a-full-heart/

207. **Rea Gostima** **22** **Died unexpectedly in her sleep** Aug, 2022
https://www.imolaoggi.it/2022/08/10/malore-improvviso-ragazza-di-22-anni-muore-nel-sonno/

208. **Mattia Ghiraldi** **22** **Died unexpectedly in his sleep** Jul, 2022
https://primacremona.it/cronaca/mattia-ghiraldi-promessa-della-motonautica-scompare-improvvisamente-a-22-anni/

209. **James Théodore** **22** **Cardiac event during game** Mar 1, 2022
https://www.eurosport.it/rugby/tragedia-in-francia-muore-in-allenamento-james-theodore-pilone-di-22-anni_sto8823863/story.shtml

210. **Caitlin Gotze** **23** **Adverse events upon vaccination** Nov 17, 2021
https://www.covidvaccineinjuries.com/covid-vaccine-stories/caitlin-gotze-23-year-old-died-while-at-work-after-employer-mandated-covid-vaccine/

211. **Antonio Salerno** **23** **Sudden death after game** Sep 12, 2022
https://www.ilrestodelcarlino.it/ferrara/cronaca/il-suo-sorriso-spento-per-sempre-in-un-attimo-e-stato-cancellato-tutto-1.8081608

212. **Abdel Rahman** **23** **Sudden death during game** Aug 3, 2021
https://sportsbeezer.com/allsports/look-an-egyptian-player-who-swallowed-his-tongue-and-died/

213. **Roy Butler** **23** **Sudden death 4 days post vaccination** Aug 13, 2021
https://thecovidblog.com/2021/08/23/roy-butler-23-year-old-irish-soccer-football-player-suffers-massive-brain-bleed-dead-four-days-after-experimental-johnson-johnson-viral-vector-dna-injection/

214. **Gilbert Kwemoi** **23** **Sudden death** Aug 14, 2021
https://www.insidethegames.biz/articles/1111716/gilbert-soet-kwemoi-dies-aged-23

215. **Riuler de Oliveira** **23** **Cardiac event** Nov 23, 2021
https://ge.globo.com/google/amp/pr/futebol/noticia/ex-athletico-e-coritiba-riuler-oliveira-morre-vitima-de-infarto-aos-23-anos.ghtml

216. **Erik Karlsson** **23** **Cardiac event** Dec 31, 2021
https://www.aftonbladet.se/sportbladet/a/MLqj85/elitlopare-dod—fick-hjartstopp-under-lopp-i-kalmar

217. **Branson King** **23** **Sudden death** Dec 11, 2021
https://www.hollywoodlanews.com/young-ice-hockey-player-dies-unexpectedly/

218. **Marin Cacic** **23** **Cardiac event** Dec 21, 2021
https://www.index.hr/sport/clanak/mladi-nogometas-nehaja-iz-senja-se-srusio-na-treningu-bore-mu-se-za-zivot/2327080.aspx

219. **Thottyanda Somanna** **23** **Cardiac event during game** Dec 25, 2021
https://www.newindianexpress.com/states/karnataka/2021/dec/25/hockey-player-dies-in-the-middle-of-a-game-inkarnataka-heart-attack-suspected-2399669.html

220. **Jamie Hoye** **23** **Sudden death** Jan 9, 2022
https://www.armaghi.com/news/lurgan-news/tributes-paid-to-young-lurgan-man-jamie-hoye-who-was-the-kindest-wee-soul/154422

221. **Michel Corbalan** **23** **Sudden death** Jan 28, 2022
https://ekstrabladet.dk/sport/anden_sport/anden_sport/dansk-europamester-doed-23-aar-gammel/9109434?ilc=c

222. **Mary Cronin** **23** **Sudden death** Apr 29, 2022
https://www.cbc.ca/news/canada/new-brunswick/mary-cronin-fredericton-soccer-unb-st-thomas-covid-19-1.6438272

223. **Michael Morgan** **23** **Sudden death** Jul 4, 2022
https://www.belfastlive.co.uk/news/northern-ireland/coalisland-community-stunned-after-sudden-24394381

224. **Amelia Smith** **23** **Died unexpectedly in her sleep** Nov 14, 2021
https://www.dailymail.co.uk/news/article-10222507/Beautiful-kind-mother-two-23-dies-suddenly-sleep-month-giving-birth.html

225. **Krzysztof Pańka** **23** **Sudden death** Nov 28, 2021
https://sport.wprost.pl/10555684/krzysztof-panka-nie-zyje-klub-potwierdzil-informacje-o-smierci-23-letniego-reprezentanta-polski.html

226. **Abraham Sié** **23** **Sudden death** Apr 6, 2022
https://www.sport-ivoire.ci/basketball/décès-dabraham-sié-la-bal-attristée

227. **Hannah Langer** **23** **Died unexpectedly in her sleep** Aug 27, 2022
https://highlandscurrent.org/2022/08/29/hannah-langer-1998-2022/

228. **Jessica Courtney** **23** **Sudden death** Jan, 2022
https://worldnewsera.com/news/uk/woman-23-dies-suddenly-from-mystery-undiagnosed-illness-as-family-pay-tribute/

229. **Traian Calancea** **24** **Stroke 10 days post vaccination** Oct, 2021
https://lanuovabq.it/it/morto-a-24-anni-il-giudice-ordina-indagate-sul-vaccino

230. **Icaro Da Silva** **24** **Cardiac event** Apr 30, 2021
https://portalmt.com.br/parada-cardiaca-morre-jogador-que-estava-em-periodo-de-teste-na-equipe-do-acao-s-a-f-c/

231. **Josh Downie** **24** **Cardiac event** May 10, 2021
https://www.bbc.com/news/uk-england-nottinghamshire-57058626

232. **Boris Sadecky** **24** **Cardiac event during game** Nov 3, 2021
https://usdaynews.com/celebrities/celebrity-death/boris-sadecky-death-cause/

233. **Michal Krowiak** **24** **Sudden death** Dec 1, 2021
https://wmeritum.pl/michal-krowiak-nie-zyje-pilkarz-mial-zaledwie-24-lata/363554

234. **Ahmed Amin** **24** **Cardiac event in locker room** Dec 22, 2021
https://www.kingfut.com/2021/12/23/third-divisions-rabat-anwar-goalkeeper-dies-of-cardiac-arrest/

235. **Dillon Quirke** **24** **Sudden death during game** Aug 5, 2022
https://www.thetimes.co.uk/article/dillon-quirke-tipperary-hurler-had-spoken-of-his-hearts-fatal-flaw-5r7k829kr

236. **Carla Steytler** **24** **Sudden death on campus** Feb, 2022
https://www.thesouthafrican.com/news/student-who-collapsed-and-died-on-bloem-campus-described-as-go-getter-breaking/

237. **Kim Lockwood** **24** **Vaccine-induced death, per coroner** Mar 24, 2022
https://www.bbc.com/news/uk-england-south-yorkshire-60757293

238. **Joyce Culla** **24** **Brain aneurysm weeks after booster** Apr 1, 2022
https://community.covidvaccineinjuries.com/joyce-culla-24-year-old-nurse-and-tiktok-star-dies-after-ruptured-brain-aneurysm-following-astrazeneca-vaccine/

239. **Elia Fiorio** **24** **Sudden death** Aug 30, 2022
https://euroweeklynews.com/2022/09/02/italian-man-dies-suddenly-unexpectedly-spain-mallorca/

240. **Finley Scholefield** **24** **Died unexpectedly in his sleep** Mar 21, 2022
https://www.theguardian.com/technology/2022/apr/28/finley-scholefield-obituary

241. **Matt Rodrigopulle** **24** **Sudden death** Sep 4, 2022
https://globalnews.ca/video/9118861/matthew-rodrigopulle-memorial/

242. **Sam Polledri** **24** **Cardiac event** Mar 2, 2022
https://www.mirror.co.uk/sport/rugby-union/polledri-gloucester-italy-death-rugby-26366673

243. **Unnamed Person** **25** **Sudden death during training** Mar 2, 2022
https://www.straitstimes.com/singapore/nsman-25-dies-after-collapsing-during-hpb-exercise-session

244. **Giacomo Gorenszach 25 Sudden death** Jun 2, 2022
https://www.udinetoday.it/cronaca/morto-giacomo-gorenszach-san-pietro-natisone.html

245. **Ashley Hipper 25 Sudden death** Jan 12, 2022
https://dailyvoice.com/new-jersey/sussex/obituaries/beloved-sussex-county-hs-grad-bio-technician-ashley-hipper-dies-suddenly-at-25/824599/

246. **P Lerkchaleampote 25 Died unexpectedly in his sleep** Mar 23, 2022
https://www.straitstimes.com/life/entertainment/thai-actor-beam-papangkorn-lerkchaleampote-dies-suddenly-in-sleep-at-25

247. **Fancis Perron 25 Sudden death after game** Sep 18, 2021
https://ottawasun.com/sports/football/tragedy-for-gee-gees-defensive-lineman-francis-perron-dies-after-game-in-toronto

248. **Michelle de Vecchi 25 Cardiac event during run** Nov 17, 2021
https://corrieredelveneto.corriere.it/treviso/cronaca/21_novembre_17/treviso-muore-25-anni-facendo-jogging-un-amico-5e8e8e8c-47ef-11ec-a5cc-cbd997036243.shtml

249. **N Mirosavljevic 25 Cardiac event** Dec 24, 2021
https://sportal.blic.rs/fudbal/domaci-fudbal/preminuo-nemanja-mirosavljevic-od-posledica-srcanog-udara/2022041920343298399

250. **Marcos Menaldo 25 Cardiac event during training** Jan 3, 2022
https://www.dailystar.co.uk/sport/football/marcos-menaldo-footballer-dies-25-25847436

251. **Abel Wasan 25 Found dead day post vaccination** Aug 1, 2021
https://rupreparing.com/news/2021/11/25/abel-wasam-25-year-old-programmer-dies-1-day-after-receiving-the-astrazeneca-covid-19-vaccine-family-seeks-answers

252. **Desiree Penrod 25 Sudden death days post vaccination** Mar 17, 2021
https://thecovidblog.com/2021/03/22/desiree-penrod-25-year-old-connecticut-educator-dead-one-week-after-johnson-johnson-viral-vector-shot/

253. **Olivia Quan 25 Brain bleed** Jul 1, 2022
https://euroweeklynews.com/2022/07/10/cause-death-recording-engineer-olivia-quan-died-suddenly/

254. **Sofia Constantino 25 Sudden death** Jan 27, 2022
https://www.direttasicilia.it/2022/01/27/sofia-costantino-termini-imerese-tragedia/

255. **Jordan Fitzgerald 25 Sudden death** Jul, 2022
https://www.thesun.ie/sport/9055587/jordan-fitzgerald-limerick-fan-died-croke-park-bus/

256. **Michele Gironella 25 Cardiac event during game** Aug 17, 2022
https://www.ilrestodelcarlino.it/macerata/cronaca/michele-gironella-1.7994646

257. **Caleb Swanigan 25 Sudden death** Jun 20, 2022
https://ftw.usatoday.com/lists/caleb-swanigan-death-25-years-old-purdue-portland

258. **Debajyoti Ghosh 25 Cardiac event during game** Mar 19, 2022
https://www.news9live.com/sports/football/shades-of-cristiano-junior-as-east-bengal-bound-footballer-debojyoti-ghosh-dies-on-field-160078

259. **Nathan Bellshaw 25 Died unexpectedly in his sleep** Apr 18, 2022
https://www.thecourier.co.uk/fp/news/dundee/3293057/nathan-bellshaw-death-dundee/

Appendix #3

Cheat Sheet for Ask Your Doctor

Is it true that thimerosal (ethylmercury) has been removed from all childhood vaccines?

No. [Note: Right before going to print with this book, it's been announced that mercury will be removed from all vaccines. Thanks to HHS Secretary Kennedy.]

Are there any vaccines still given to children that contain thimerosal?

Yes.

Which vaccine products contain thimerosal, in case I want to avoid those?

Despite decades of being misled that mercury is no longer in vaccines, it has remained in six popular vaccine products still given to American children:

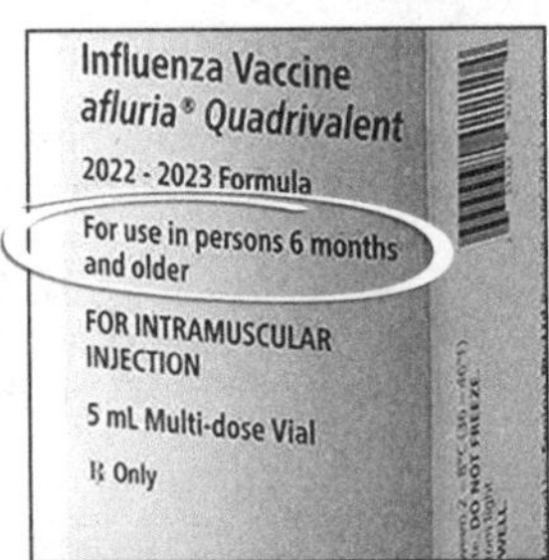

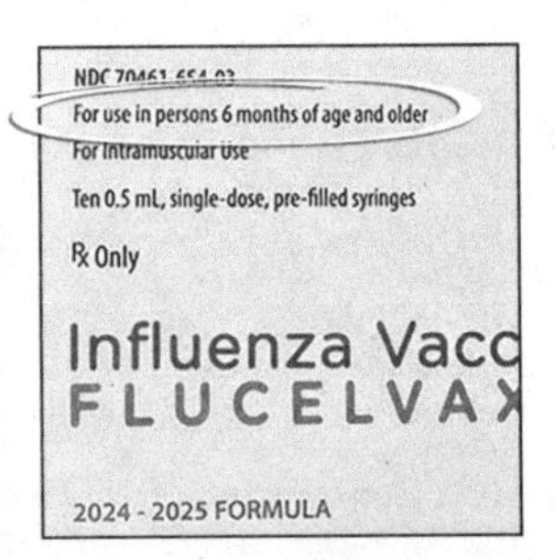

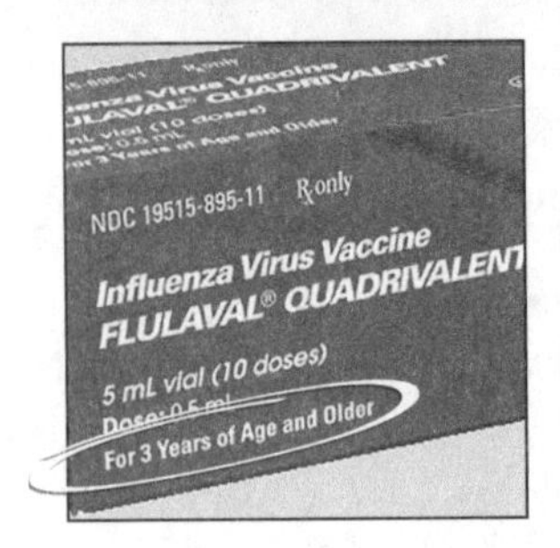

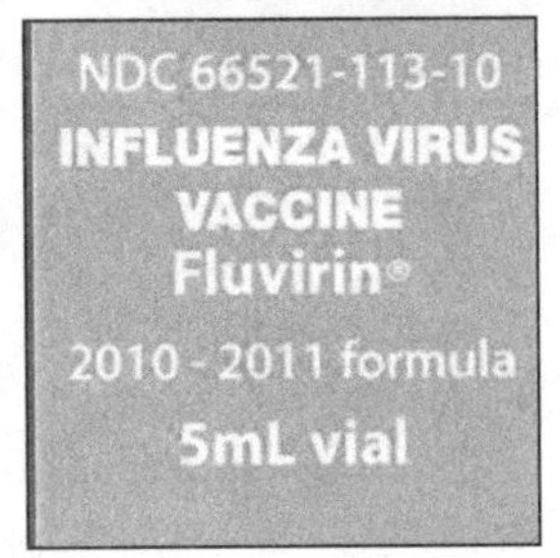

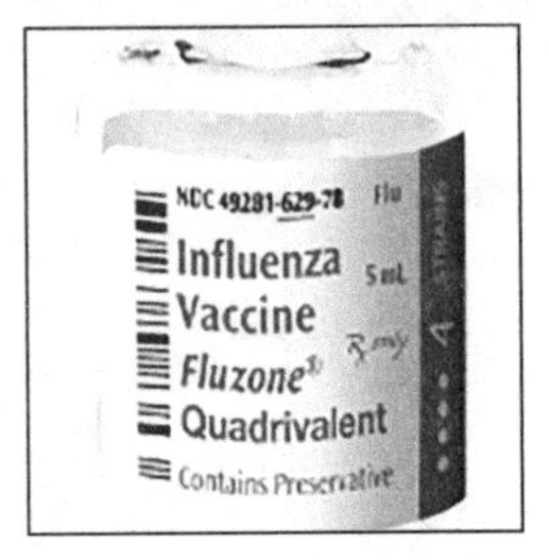

Do childhood vaccine package inserts acknowledge any neurological injuries as possible adverse events from their vaccines?

Yes, and the list is extensive.

Which of these ingredients are contained in vaccines given to children? Gelatin from boiled pig skin, Chicken embryo protein, Blood from the hearts of cow fetuses, Human fetus DNA fragments, Albumin from human blood plasma, Oil extracted from shark livers, proteins from worm ovaries, Monkey kidney DNA fragments, Formaldehyde, Polysorbate 80, Potassium chloride, Phenol, Borax/sodium borate, Monosodium glutamate (MSG), Aluminum salts, Thimerosal/ethylmercury, Triton X-100

All.

If a child is injured by a vaccine, can the family sue the vaccine manufacturer?

No.

If a child is injured by a vaccine, what government agency compensates the family?

The National Vaccine Injury Compensation Program is run by the Health Resources & Services Administration (HRSA), an agency within the Department of Health & Human Services (HHS).

Has the US Government ever compensated any family for autism caused by a vaccine?

Yes. See Dr. Jon Poling, Chapter Eight.

Has the US Government ever compensated any family for any neurological injuries or brain damage caused by a vaccine?

Yes, on many, many occasions. See Chapter Four.

Can the BCG vaccine be beneficial to health beyond protection from tuberculosis?

Yes. See Chapter Eleven.

What is the main adverse effect associated with human exposure to mercury?

When mercury damages the human body, the main adverse effects are neurological, developmental, cognitive and behavioral, including delayed speech and loss of acquired skills. You know, a lot like severe autism. See Chapter Nine.

What do you believe causes autism?

If we think in terms of brain damage, learning and speech delay, neurodevelopmental problems, and neurological injury, the vaccine-makers' own package inserts acknowledge these neurological side-effects from their products:

MMR Vaccine
Seizures

Encephalomyelitis
Transverse myelitis
Syncope
Polyneuropathy
Ataxia
Guillain-Barré Syndrome
Progressive Neurological Disorder

Varicella Vaccine (chickenpox)
Ataxia
Encephalitis
Transverse myelitis
Guillain-Barré syndrome
Seizures
Bell's Palsy
Stroke
Meningitis

Hepatitis B Vaccine
Multiple sclerosis

Influenza Vaccine
Guillain-Barré syndrome
Seizures

Pneumococcal Conjugate Vaccine
Seizures

Oral Polio Vaccine
Vaccine-associated paralytic poliomyelitis

Meningococcal Vaccine
Guillain-Barré syndrome

HPV Vaccine
Guillain-Barré syndrome

DTP/DTaP Vaccine (Note: DTaP currently used)
Seizures
Prolonged convulsions
Encephalopathy
Neuropathy
Guillain-Barré syndrome

Hypotonic-hyporesponsive episodes
Lowered consciousness
Persistent neurologic symptoms
Unresponsiveness
Coma
Progressive neurologic disorders

Covid-19 Vaccines
Hemorrhagic Stroke
Guillain-Barré syndrome
Transverse myelitis
Encephalitis
Meningitis
Bell's Palsy
Seizures
Convulsive Disorders

Note: The list above is far from an account of all acknowledged vaccine side-effects. It's a list of only those side-effects that are neurological (many of which can have crossover with autism).

Appendix #4: News You Likely Missed

NEWS

State Surgeon General Warns Young Men COVID Vaccines Pose 'High Risk' of Death

BY FATMA KHALED ON 10/8/22 AT 5:46 PM EDT

Florida's Surgeon General Joseph A. Ladapo warned on Friday against young men receiving COVID-19 vaccines, citing a disputed analysis by the state health department that they pose an "abnormally high risk" of death.

THE WALL STREET JOURNAL.

Are Covid Vaccines Riskier Than Advertised?

There are concerning trends on blood clots and low platelets, not that the authorities will tell you.

By Joseph A. Ladapo and Harvey A. Risch
June 22, 2021 1:09 pm ET

One remarkable aspect of the Covid-19 pandemic has been how often unpopular scientific ideas, from the lab-leak theory to the efficacy of masks, were initially dismissed, even ridiculed, only to resurface later in mainstream thinking. Differences of opinion have sometimes been rooted in disagreement over the underlying science. But the more common motivation has been political.

REUTERS

4 minute read · November 18, 2021 1:31 PM PST · Last Updated a year ago

Wait what? FDA wants 55 years to process FOIA request over vaccine data

(Reuters) - Freedom of Information Act requests are rarely speedy, but when a group of scientists asked the federal government to share the data it relied upon in licensing Pfizer's COVID-19 vaccine, the response went beyond typical bureaucratic foot-dragging.

As in 55 years beyond.

Daily Mail .com

Up to one in 7,000 American teens suffered heart inflammation after their Covid vaccine, study suggests

- Scientists at Kaiser Permanente reviewed 340 cases out of 6.9million vacinees
- Found those getting a second dose were most likely to suffer myocarditis
- This struck within the first seven days in the vast majority of cases

By LUKE ANDREWS HEALTH REPORTER FOR DAILYMAIL.COM
PUBLISHED: 17:02 EDT, 3 October 2022 | UPDATED: 17:11 EDT, 3 October 2022

Thousands of American teenagers may have suffered heart inflammation after getting a Covid jab, a study suggests.

Researchers found up to one in 7,000 boys aged 12 to 15 years old developed myocarditis after receiving the Pfizer vaccine.

Pfizer Exec Concedes COVID-19 Vaccine Was Not Tested on Preventing Transmission Before Release

By Jack Phillips | October 11, 2022 Updated: October 13, 2022
A A Print

A Pfizer executive said Oct. 10 that neither she nor other Pfizer officials knew whether its COVID-19 vaccine would stop transmission before entering the market last year.

Daily Mail .com

Revealed: PR firm that represents Pfizer and Moderna also sits on CDC vaccine division - sparking major conflict of interest concerns

PUBLISHED: 12:16 EDT, 12 October 2022 | UPDATED: 14:27 EDT, 12 October 2022

Autopsy Confirms NY State College Student Died From "COVID-19 Vaccine-Related Myocarditis"

(TJV NEWS) 24-year-old New York college student George Watts Jr. died on October 27, 2021, due to complications related to the Pfizer Covid-19 shots he took in August and September.

It was revealed recently that the Bradford County Coroner's Office listed the COVID vaccine as the cause of death.

The New York Times

Johns Hopkins Scientist: 'A Medical Certainty' Pfizer Vaccine Caused Death of Florida Doctor

Dr. Jerry L. Spivak, an expert on blood disorders at Johns Hopkins University, told the New York Times Tuesday that he believes "it is a medical certainty" that Pfizer's COVID vaccine caused the death of Dr. Gregory Michael.

According to the New York Times:

> "Dr. Jerry L. Spivak, an expert on blood disorders at Johns Hopkins University, who was not involved in Dr. Michael's care, said that based on Ms. Neckelmann's description, 'I think it is a medical certainty that the vaccine was related.'

"We don't believe at this time that there is any direct connection to the vaccine."

— Pfizer

Volume 40, Issue 40, 22 September 2022, Pages 5798-5805

Serious adverse events of special interest following mRNA COVID-19 vaccination in randomized trials in adults

3.4. Harm-benefit considerations

In the Moderna trial, the excess risk of serious AESIs (15.1 per 10,000 participants) was higher than the risk reduction for COVID-19 hospitalization relative to the placebo group (6.4 per 10,000 participants). [3] **In the Pfizer trial,** the excess risk of serious AESIs (10.1 per 10,000) was higher than the risk reduction for COVID-19 hospitalization relative to the placebo group (2.3 per 10,000 participants).

Massachusetts Death Certificates Show Excess Mortality Could be Linked to COVID Vaccines

By Madhava Setty
Global Research, November 23, 2022

Actuaries raise alarm that Australians are unexpectedly dying at incredibly high rate

BY RHODA WILSON ON DECEMBER 16, 2022 • (67 COMMENTS)

The Australian government should be urgently investigating the "incredibly high" 13% excess death rate in 2022, the country's peak actuarial body says.

Why are so many footballers collapsing? There has been a worryin spike in cardiac arrests and stars retiring with heart-related issues, but leading sports cardiologist insists it is NOT to do with Covid vaccine

PUBLISHED: 17:34 EDT, 15 December 2021 | UPDATED: 19:38 EDT, 15 December 2021

Nine years separated Marc-Vivien Foe's death from a heart attack on a pitch in Ly to Fabrice Muamba's near-fatal collapse at Tottenham, and another nine passed before Christian Eriksen was brought back to life at Euro 2020 this summer.

Three of the most harrowing in-game episodes that football has seen were sprea over more than 18 years, yet it feels that barely a week goes by at the moment without news of another cardiac-related incident in the game.

THE DAILY SCEPTIC

Covid Vaccines Up to 100 Times More Likely to Cause Serious Injury to a Young Adult Than Prevent It, Say Top Scientists

BY WILL JONES 7 SEPTEMBER 2022 7:00 AM

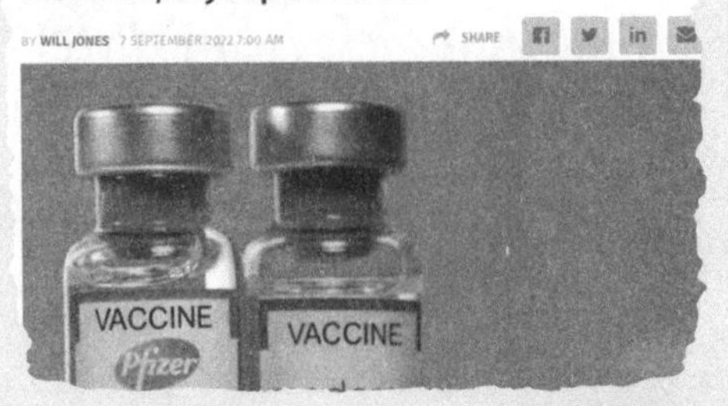

FOX NEWS

FOX NEWS FLASH · Published July 22, 2022 7:29pm EDT

Dr. Deborah Birx says she 'knew' COVID vaccines would not 'protect against infection'

The former White House COVID response coordinator downplays vaccine efficacy

CNN health

FDA vaccine advisers 'disappointed' and 'angry' that early data about new Covid-19 booster shot wasn't presented for review last year

Updated 9:24 AM EST, Wed January 11, 2023

"I was angry to find out that there was data that was relevant to our decision that we didn't get to see," said Dr. Paul Offit, a member of the Vaccines and Related Biological Products Advisory Committee, a group of external advisers that helps the FDA make vaccine decisions. "Decisions that are made for the public have to be made based on all available information – not just some information, but all information."

Science

SCIENCEINSIDER | PLANTS & ANIMALS

FDA no longer requires animal testing before human drug trials

10 JAN 2023

New medicines need not be tested in animals to receive U.S. Food and Drug Administration (FDA) approval,

Report reveals Pfizer shot caused avalanche of miscarriages, stillborn babies

Among the first reports handed over by Pfizer was a 'Cumulative Analysis of Post-authorization Adverse Event Reports' describing events reported to Pfizer up until February 2021.

Pfizer's report states that there were 23 spontaneous abortions (miscarriages), two premature births with neonatal death, two spontaneous abortions with intrauterine death, one spontaneous abortion with neonatal death, and one pregnancy with "normal outcome." That means that of 32 pregnancies with known outcome, 28 resulted in fetal death.

Babies Dying In Scotland: Govt Issues Investigation Into Rising Infant Mortality Rate

Written by Kashish Sharma | Updated : October 4, 2022 10:49 AM IST

The Scottish government has issued an investigation into the suddenly rising infant mortality rate in the country. There have been two spikes over a period of six months. Earlier the figures had shown that the death rate for babies under one year of age was at its highest in 10 years. The increase was high enough to initiate an investigation.

LIFESTYLE > HEALTH

New Study Shows COVID-19 Vaccine Does Cause Changes to People's Menstrual Cycles

The new study confirms findings that there is a link between vaccination against COVID-19 and an average increase in menstrual cycle length

By Amanda Taylor | Published on September 28, 2022 09:13 PM

ISRAEL NATIONAL NEWS

New study: COVID vaccination shown to decrease sperm counts

Effect begins 2 months following vaccination & persists for at least 5 months, when study ended.

Y Rabinovitz

Jun 22, 2022, 3:51 PM (GMT+3)

vaccine vaccination Coronavirus

A new study conducted by Israeli researchers at Shamir Medical Center, Tel Aviv University, Herzliya Medical Center, and Sheba Medical Center, has published disturbing findings regarding the possible impact of COVID vaccination on male fertility.

CTV NEWS

CALGARY | News

Deaths with unknown causes now Alberta's top killer: province

Published July 5, 2022 6:16 p.m. ET
Updated July 6, 2022 8:40 p.m. ET

By Nicole Di Donato
CTV News Calgary Multimedia Journalist

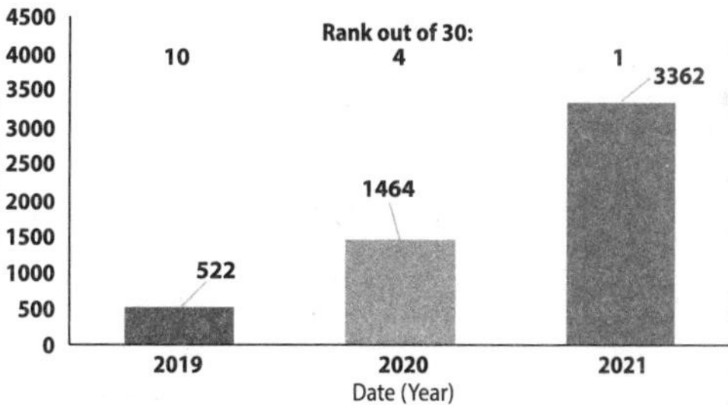

EuroWeekly NEWS
NEWS AND VIEWS SINCE 1998

Pilot of Boeing flight to St. Petersburg dies suddenly on board plane

By Matthew Roscoe · 19 September 2022 · 16:08

Pilot Dies Suddenly After Takeoff from Chicago Airport

By Richard Moorhead

November 27, 2022 at 8:41am

An American Airlines flight was forced to return to its take-off location Nov. 26 after the aircraft's pilot suffered a medical emergency in-flight.

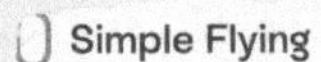

Home › Airline News › Biman Bangladesh Pilot Dies After Inflight Medical Er

Pilot Dies After Inflight Medical Emergency

BY JAKE HARDIMAN PUBLISHED SEP 5, 2021

Photo: Md Shaifuzzaman Ayon via Wikimedia Common

INDEPENDENT

Pilot dies in bathroom on Miami flight carrying 271 passengers

August 17, 2023

Two Indian Pilots Die In A Day, One Collapsed Moments Before Flying, Another Mid-Air

The DGCA is investigating both incidents to determine the cause of the cardiac arrests.

Thu, 17 Aug 2023

Daily Mail .com

Veteran British Airways pilot dies after suffering heart attack in hotel shortly before he was due to captain flight from Cairo to Heathrow

PUBLISHED: 06:21 EDT, 12 March 2023

USA TODAY

Alabama high school basketball star Caleb White dies after collapsing during pickup game

Aug. 12, 2023

Daily Mail .com

World's sixth best young water-skier 'dies suddenly' aged just 18

PUBLISHED: 18:19 EDT, 18 May 2023

Mail Online

Two-time national champion boxer Jude Moore dies suddenly aged just 19

PUBLISHED: 06:14 EDT, 21 March 2023

Devastated family pay tribute to son, 15, who died during football match

15 Mar 2023

WSAZ NewsChannel 3

High school senior dies after suffering heart attack at school, officials say

Jan. 10, 2023

NEWS

Dominican basketball player who previously blamed COVID vaccine for rare heart condition dies of heart attack

Published June 24, 2023 7:32pm EDT

N Q Report

National Guard Soldier Suffers TWO Heart Attacks After Moderna "Vaccine"

April 9, 2023

Family heartbreak as daughter suddenly collapses and dies in dad's arms with 'heart attack'

1 Mar 2023

ILY NEWS

ollege basketball star Derek Gr ead at 20 after collapsing at mpus basketball camp

ami Ganz · New York · Jul 30, 2022 at 4:57 pm

AD NEWS REGION SPORT SHOW PLAY PODCAST PUZZL

Wouter (14) collapses on hockey field and dies, school and club in mourning

Sebastian Quekel 08-12-21, 13:49 Last update: 08-12-21, 15:58

NEWS SPORT SHOWBIZ NINA NEAR VIDEO PUZZLE

Football player Niels De Wolf (27) died after being struck by heart failure after a game on Sunday

Kristof Pieters 07-10-21, 08:28 Last update: 07-10-21, 09:12

Appendix #5

Researching the Research

A superb paper by Dr. Toby Rogers provides a detailed and expansive look at hundreds of studies seeking or claiming to seek answers about causes of autism. Almost all the studies must be disqualified because they did not include vaccine products among the possible toxicants to be considered. (If you don't look, you won't find.) Other studies are disqualified due to financial or other conflicts such as research is funded by regulators who recommended mass vaccination, key study participants funded by Pharma, key study participants are paid consultants to Pharma companies, researchers or regulators auditioning for Pharma jobs, etc. And as in any field, some of the studies have fatal flaws in design or method.

Rogers writes, "The holy grail in autism research is to find vaccinated vs. unvaccinated studies. Thankfully there are now six good studies that we can rely on." [1]

1. Gallagher and Goodman (2008) found that boys who received all three doses of hepatitis B vaccine were 8.63 times more likely to have a developmental disability, including autism, than boys who did not receive all three doses. [2]

2. Gallagher and Goodman (2010) found that boys "who received the first dose of hepatitis B vaccine during the first month of life had 3-fold greater odds for autism diagnosis" as compared with "boys either vaccinated later or not at all" (p. 1669). [3]

    (The study above is about the effects of just one injection. No one knows the effect of injecting all 77 recommended vaccine doses combined over time.)

3. Anthony Mawson is an epidemiologist and professor with a long track record of published research, including in *The Lancet*. In 2017, Mawson and his co-authors studied more than 650 children between the ages of 6 and 12. More than 400 were vaccinated, and 261 were unvaccinated.

    Vaccinated children in the study were almost 5 times more likely to be diagnosed with autism than unvaccinated children. Of note, the

1  2  3 

vaccinated children were also significantly more likely than the unvaccinated to have been diagnosed with

- a learning disability (5.7% vs. 1.2%)
- ADHD (4.7% vs. 1.0%)
- any neurodevelopmental disorder, e.g., learning disability, ADHD or autism (10.5% vs. 3.1%)
- any chronic illness (44.0% vs. 25.0%) [1]

4. Mawson, Bhuiyan, Jacob, and Ray (2017) conducted a separate analysis on data related to preterm children (aka "premies"), and found:
    - No association between preterm birth and neurodevelopmental disability [learning disability, ADHD, and/or autism] in the absence of vaccination.
    - Preterm birth coupled with vaccination increased the odds of neurodevelopmental disability by more than twelve-fold compared to preterm birth without vaccination. [2]
5. Hooker and Miller (2021) compared vaccinated children to unvaccinated children for the incidence of several health conditions including autism. Children who were "vaccinated and not breastfed" had a more than 12-fold higher risk of autism. Children who were "vaccinated and delivered via cesarean section" had a more than 18-fold higher risk of autism. [3]
6. Mawson and Jacob (2025) studied 47,155 children born in the Florida State Medicaid program and continuously enrolled until age 9. The analysis revealed that:
    - vaccination was associated with significantly increased odds for all measured neurodevelopmental disorders;
    - among children born preterm and vaccinated, 39.9% were diagnosed with at least one neurodevelopmental disability compared to 15.7% among those born preterm and unvaccinated; and
    - the relative risk of autism spectrum disorder increased according to the number of medical visits that included vaccinations.

---

1 
2 
3 

> Children with just one vaccination visit were 1.7 times more likely to have been diagnosed with autism than the unvaccinated, whereas those with 11 or more visits that included vaccinations were 4.4 times more likely to have been diagnosed with autism than those with no visit for vaccination. [1]

Last Note in this book: Research from Thomas and Margulis (2016) shows that the autism rate in children with no vaccines is 1 in 715, and the autism rate in vaccinated children is 1 in 31. [2]

---

1 2